B31

THE NEW ILLUSTRATED GUIDE TO

MODERN
TANKS
& FIGHTING VEHICLES

a Salamander book

Published by Salamander Books Limited
LONDON ● NEW YORK

THE NEW ILLUSTRATED GUIDE TO
MODERN
TANKS
& FIGHTING VEHICLES

DAVID MILLER

A Salamander Book

Published by
Salamander Books Ltd.,
129-137 York Way,
London N7 9LG,
United Kingdom

©Salamander Books Ltd. 1992

ISBN 0-86101-682-3

Distributed in the United Kingdom by
Hodder & Stoughton Services, PO Box
6, Mill Road, Dunton Green, Sevenoaks,
Kent TN13 2XX.

All correspondence concerning the
content of this volume should be
addressed to the publisher.

Contents

Combat vehicles are arranged alphabetically within national groups.

	PAGE
ARGENTINA	
TAM Medium Tank	6
AUSTRIA	
Panzerjäger K 4KH7FA SK 105	
Light Tank/Tank Destroyer	8
BRAZIL	
X1A2 Medium Tank	10
CHINA	
Type 59 Main Battle Tank	12
Type 69 Main Battle Tank	14
FRANCE	
AMX-13 Light Tank	16
AMX-30B2 Main Battle Tank	20
MK F3 155mm Self-propelled	
Howitzer	26
AU 1 (GCT) 155mm	
Self-propelled Gun	28
Leclerc Main Battle Tank	32
GERMANY	
Jagdpanzer 4-5 Kanone, Jaguar	
1 Rakete and Jaguar 2 Tank	
Destroyers	34
Leopard 1 Main Battle Tank	38
Leopard 2 Main Battle Tank	42
ISRAEL	
Merkava Main Battle Tank	48
Soltam L-33 155mm Self-propelled	
Gun/Howitzer	54
ITALY	
C1 Ariete Main Battle Tank	56
JAPAN	
Type 61 Main Battle Tank	58
Type 74 Main Battle Tank	60
Type 75 155mm Self-propelled	
Howitzer	64
SOUTH KOREA	
Type 88 Main Battle Tank	66
SOVIET UNION	
SO-122 (2S1) Gvodzika (M-1974)	
122mm Self-propelled Howitzer	70

Credits

Author: David Miller is an ex-officer in the British Army who now writes as a full-time author on military subjects.

Editor: Bob Munro

Designer: Phil Gorton

Colour drawings: ©Salamander Books Ltd., and ©Profile Publications Ltd.

Filmset by The Old Mill

Colour reproduction by Scantrans Pte.

Photographs: The publishers wish to thank all the official international governmental archives, weapons systems manufacturers and private collections who have supplied photographs for this book.

Printed in Hong Kong

SO-152 (2S3) Akatsiya (M-1973) 152mm Self-propelled Gun/Howitzer   72
SM-240 (2S4) Tyulpan (M-1975) 240mm Self-propelled Mortar   74
SO-203 (2S7) (M-1975) 203mm Self-propelled Gun   76
ASU-85 Self-propelled Anti-tank Gun   78
PT-76 Light Amphibious Tank   82
SO-120 (2S9) 120mm Self-propelled Howitzer/Mortar   84
T-55 Main Battle Tank   86
T-62 Main Battle Tank   88
T-64 Main Battle Tank   90
T-72 Main Battle Tank   92
T-80 Main Battle Tank   96

SWEDEN
Bandkanon 1A 155mm Self-propelled Gun   98
Infanterikanonvagn 91 Light Tank/Tank Destroyer   102

Stridsvagn Strv-103 Main Battle Tank   106

SWITZERLAND
Pz 61 and 68 Main Battle Tanks   108

UNITED KINGDOM
AS-90 155mm Self-propelled Gun   110
Centurion Main Battle Tank   114
Challenger Main Battle Tank   116
Chieftain Main Battle Tank   122
Vickers Main Battle Tank   126

UNITED STATES
M1 Abrams Main Battle Tank   130
M41 Walker Bulldog Light Tank   138
M47 Medium Tank   140
M48A5 Medium Tank   142
M60A3 Main Battle Tank   144
M107 175mm Self-propelled Gun/M110A2 203mm Self-propelled Howitzer   148
M109 155mm Self-propelled Howitzer   152
M551 Sheridan Light Tank   156
Commando Stingray Light Tank   158

TAM Medium Tank

Countries of origin: Germany (design), Argentina (production).
Crew: 4.
Armament: One 105mm gun; one 7.62mm machine-gun co-axial with main armament; one 7.62mm anti-aircraft machine-gun; eight smoke dischargers.
Armour: Classified.
Dimensions: Length (with gun forwards) 27ft (8.23m); length (hull) 22ft 3in (6.78m); width 10ft 8in (3.25m); height 7ft 11in (2.42m).
Weight: Combat 67,252lb (30,500kg).
Ground pressure: 11.23lb/in² (0.79kg/cm²).
Engine: MTU six-cylinder diesel developing 710hp at 2,200rpm.
Performance: Road speed 46mph (75km/h); road range 342 miles (550km); vertical obstacle 3ft 3in (1m); trench 8ft 2in (2.5m); gradient 60 per cent.
History: Production complete. In service with Argentinian Army (350).

The Argentinian Army has in the past obtained most of its equipment from the United States but recent American policy has led to a drastic curtailment in the supply of arms to many countries, especially those in South America. So in 1974 the Argentinian Army placed a contract with the West German company of Thyssen Henschel for the design and development of the Tanque Argentino Mediano (TAM) medium tank, and a contract was placed at the same time for the design and development of an infantry fighting vehicle to operate with the TAM, called the Véhiculo Combate Infanteria (VCI). Under the terms of the contract three prototypes of both the TAM and the VCI were to be supplied and a factory was to be established in Argentina to undertake production of both vehicles, which would initially be assembled from components supplied from West Germany but in time would be mostly manufactured in Argentina, not only providing some employment but also saving the country valuable foreign exchange costs.

Both the TAM and the VCI are based to a large extent on the chassis of the Marder Mechanised Infantry Combat Vehicle which entered service with the West German Army in 1971. The hull of the TAM is of all-welded steel construction with the driver seated at the front of the well-sloped hull on the left with the engine to his right. The all-welded turret is mounted at the rear of the hull with the commander and gunner on the right and the loader on the left. The suspension system is of the torsion bar type and consists of six dual rubber-tyred road wheels with the drive sprocket at the front, idler at the rear and three track return rollers.

Above: The TAM Medium Tank has an operational range of 342 miles (550km) which can be increased to 559 miles (900km) by fitting two long-range auxiliary fuel tanks to the rear of the hull. The TAM is essentially a Marder Mechanised Infantry Combat Vehicle hull with a new turret and a 105mm gun at the rear of the hull.

Above: The TAM Medium Tank was originally developed by the German company Thyssen Henschel to meet the requirements of the Argentinian Army. 350 were acquired for service.

The first, second, fifth and sixth road wheel stations are provided with a hydraulic shock absorber. The basic model has a range on internal fuel tanks of some 342 miles (550km) but to increase the range to 559 miles (900km) two long range fuel tanks can be mounted at the rear of the hull. The basic vehicle can ford to a depth of 4ft 7in (1.4m) without any preparation but with a snorkel fitted it can ford to a depth of 13ft 2in (4m).

Main armament consists of a 105mm gun which can fire fixed APFSDS, HEAT, HE-T, HESH and WP-T rounds, with a total of 50 rounds being carried, and loaded into the TAM via a door in the rear of the hull or via a small circular door in the left side of the turret. A 7.62mm machine-gun is mounted co-axial with the main armament and a similar weapon is mounted on the turret roof for anti-aircraft fire; four electrically-operated smoke dischargers are fitted either side of the turret. The fire control system consists of a panoramic sight for the commander which has a magnification of from x6 to x20, a coincidence rangefinder which is also operated by the commander, while the gunner is provided with a sight with a magnification of x8.

Below: The TAM Medium Tank sports a well-sloped glacis plate, leading up and back to a turret which extends almost to the rear of the hull. Visible in this view is the external mantlet which houses the tank's 105mm main gun.

Panzerjäger K 4KH7FA SK105 Light Tank/Tank Destroyer

Country of origin: Austria.
Crew: 3.
Armament: One 105mm gun; one 7.62mm machine-gun co-axial with main armament; three smoke dischargers either side of turret.
Armour: 0.4in-1.6in (10mm-40mm).
Dimensions: Length (with gun forwards) 25ft 6in (7.76m); length (hull) 18ft 3in (5.58m); width 8ft 2in (2.5m); height 8ft 2in (2.51m).
Weight: Combat 38,587lb (17,500kg).
Ground pressure: 9.67lb/in² (0.68kg/cm²).
Engine: Steyr 7FA turbo-charged 6-cylinder diesel developing 320hp at 2,300rpm.
Performance: Road speed 40.4mph (65km/h); range 323 miles (520km); vertical obstacle 2ft 8in (0.8m); trench 7ft 11in (2.41m); gradient 75 per cent.
History: Entered service with Austrian Army in 1973. Now in service with following armies: Argentina (150), Austria (250), Bolivia (34), Morocco (109), Tunisia (54).

In 1965 Saurer-Werke commenced the development of this well-armed and highly mobile tank destroyer to meet the requirements of the Austrian Army. The chassis uses many components of an earlier range of APCs but its layout is quite different with the driver's compartment at the front, turret in the centre and the engine and transmission at the rear. The hull is of all-welded construction and provides the crew with protection from small arms fire and shell splinters. The suspension is of the torsion bar type and consists of five dual rubber-tyred road wheels with the drive sprocket at the rear, idler at the front and three track return rollers. The first and last road wheel stations have an hydraulic shock absorber.

The FL-12 turret is made under licence in Austria from the French company Fives-Lille-Cail and is identical to that fitted to the AMX-13 light tank and the Brazilian EE-17 (6x6) tank destroyer. This turret is of the oscillating type with the 105mm gun fixed in the upper half which in turn pivots on the lower part. The gun can be elevated from −6° to +13° and the turret traversed through a full 360° in 12 to 15 seconds. The 105mm gun is fed from two revolver-type magazines in the turret bustle, each of which holds six rounds of ammunition. Empty cartridge cases are ejected outside of the turret through a small trap door in the turret rear. The two magazines have enabled the crew to be reduced to

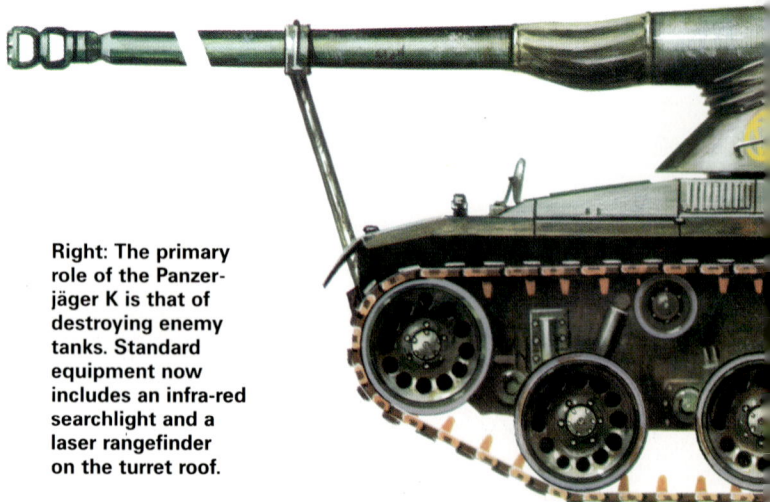

Right: The primary role of the Panzerjäger K is that of destroying enemy tanks. Standard equipment now includes an infra-red searchlight and a laser rangefinder on the turret roof.

Above: A Panzerjäger K firing its 105mm gun. Empty cartridge cases are ejected through a door in the rear of the turret.

three men — commander, gunner and driver — and also allow a high rate of fire to be achieved for a short period; on the other hand, once the 12 rounds have been fired at least one of the crew has to leave the vehicle to carry out manual reloading of the two magazines. A total of 44 rounds of 105mm ammunition are carried, which can be a mixture of the following types: HE with the complete round weighing 41lbs (18.4kg); HEAT with the complete round weighing 39lbs (17.7kg) which will penetrate 14in (360mm) of armour at an incidence of 0° or 6in (150mm) of armour at an incidence of 65°; smoke with the complete round weighing 42lbs (19.1kg). Mounted co-axial to the right of the main armament is a 7.62mm MG42/49 machine-gun and mounted on either side of the turret are three electrically operated smoke dischargers; a total of 2,000 rounds of 7.62mm ammunition are carried. Recently most vehicles have been fitted with a laser rangefinder mounted externally on the turret roof and above this has been mounted an infra-red/white-light searchlight. The Kürassier K, as the vehicle is often called, has no NBC system and no deep fording capability.

X1A2 Medium Tank

Country of origin: Brazil.
Crew: 3.
Armament: One 90mm gun; one 7.62mm machine-gun co-axial with main armament; one 12.7mm anti-aircraft machine-gun; six smoke dischargers.
Armour: Classified.
Dimensions: Length (with armament) 23ft 3in (7.1m); length (hull) 21ft 4in (6.5m); width 8ft 6in (2.6m); height (to turret top) 8ft (2.45m).
Weight: Combat 41,895lb (19,000kg).
Ground pressure: 8.96lb/in² (0.63kg/cm²).
Engine: Scania DS-11 6-cylinder diesel developing 300hp at 2,200rpm.
Performance: Road speed 34mph (55km/h); range 466 miles (750km); vertical obstacle 2ft 3in (0.7m); trench 6ft 10in (2.1in); gradient 70 per cent.
History: Production complete. In service with the Brazilian Army (50).

The X1A2 is an entirely new tank produced for the Brazilian Army by Bernardini of São Paulo. It does incorporate features of the earlier X1A and X1A1 tanks, but these were essentially rebuilds of the American M3A1 Stuart light tank, some 200 of which were supplied by the US over 30 years ago.

The hull is of all-welded construction and is divided up into three compartments, driver's at the front, fighting in the centre and the engine at the rear. The driver is seated on the left side with ammunition stowed to his right. The two other crew members are seated in the all-welded steel turret, the commander on the

left and the gunner on the right, both with a single piece hatch cover that opens to the rear and vision devices. The engine is made under licence in Brazil and is coupled to a manual transmission with two forward and one reverse gear. The suspension is of the vertical volute type, each side having three bogies each with two road wheels and the drive sprocket at the front, idler at the rear and three tank return rollers that support the inside of the track only.

Main armament consists of a 90mm gun which has a double baffle muzzle brake; this fires a HEAT projectile weighing 8.04lbs (3.65kg) with a muzzle velocity of 831 yards/s (760m/s), which will penetrate 12in (320mm) of armour at an incidence of 0°, and an HE projectile weighing 12.56lbs (5.7kg) with a muzzle velocity of 711 yards/s (650m/s). Mounted co-axial with the main armament is a 7.62mm machine-gun and mounted on the turret roof is a 12.7mm anti-aircraft machine-gun. A total of 66 rounds of 90mm, 2,500 rounds of 7.62mm and 750 rounds of 12.7mm ammunition are carried. Three electrically operated smoke dischargers are mounted either side of the turret. Optional equipment includes the replacement of the 90mm gun with a 105mm gun, the installation of a laser rangefinder, infra-red night vision equipment and an air-conditioning system. The X1A2 has no inherent amphibious capability although it can ford to a depth of 4ft 3in (1.3m).

Below: Incorporating lessons learnt on the X1A and X1A1 medium Tanks, the X1A2 was built specifically for the Brazilian Army, with a final total of 50 being acquired. Visible is the distinctive triple-bogey, dual-road wheel arrangement, and the double baffle muzzle brake on the 90mm main gun.

Type 59 Main Battle Tank

Country of origin: China.
Crew: 4.
Armament: One Type 59 (D-10T copy) 100mm rifled gun; one Type 59T 7.62mm co-axial machine-gun; one Type 59T AA 7.62mm machine-gun.
Armour: 1.5in-8in (39mm-203mm).
Dimensions: Length (including main armament) 29ft 6in (9.00m); length (hull) 19ft 10in (6.04m); width 10ft 8in (3.27m); height 8ft 6in (2.59m).
Weight: Combat 79,300lb (36,000kg).
Ground pressure: 11.38lb/in² (0.80kg/cm²).
Engine: Model 12150L V-12 liquid-cooled diesel developing 520hp at 2,000rpm.

Performance: Road speed 31mph (50km/h); range 273 miles (440km); vertical obstacle 2ft 7in (0.79m); trench 8ft 10in (2.7m), gradient 60 per cent.
History: First production Type 59 completed in 1957; production complete. In service with: Albania (15), Bangladesh (100 approx), China (6,000), Congo (15), Kampuchea (some), North Korea (175), Pakistan (1,100), Tanzania (30), Vietnam (160), Zimbabwe (35).

When the People's Liberation Army (PLA) defeated Chiang Kai-Shek's Kuomintang in 1949 they inherited a mixture of old Japanese and American tanks to supplement their own small numbers of elderly Soviet MBTs. There was an obvious need to establish their own MBT production facilities as soon as possible and to this end a small number of T-54s were procured from the then friendly Soviet Union. This model was copied exactly and put into production at a new factory at Baotou, near Peking. Designated the Type 59 by the West, the first production versions reached the PLA in about 1957 and production rose steadily to a rate of some 500-700 per year by the 1970s, peaking at about 1,000 per year in the early 1980s. Although not officially confirmed, it would appear that production has now ended.

The original Type 59 was indentical to the Soviet Type 54, but the tank has been progressively developed during its long production run in China. Two

versions are known to exist with a 105mm gun. One has been developed in China and is designated the Type 59-II by the PLA and the Type 59M1984 by the US Army. This is fitted with a 105mm gun with a thermal sleeve and fume extractor, which from visual inspection looks very similar to the British L7. The other version has been produced in the United Kingdom by Royal Ordnance and mounts a standard L7A3 105mm gun, and also sports additional armour. This MBT was tested in Pakistan in 1987, but no order was forthcoming.

The producers of the tank, the Chinese North Industries Corporation, offer no less than eight separate packages to enable current users of the Type 59 or T-54/-55 to upgrade their fleet. These enhancements include everything from appliqué armour, through new optical and electronic devices to either a new-model 100mm gun or a new Chinese-developed 105mm gun.

Above: T-59 MBTs of the PLA move into action during a live-fire exercise. Based on the Soviet T-54 MBT, the Type 59 has provided the backbone of the PLA's tank force for many years, as well as winning a healthy number of export orders.

Left: Though the ancestry of the T-59 can be traced back to the early 1950s, when numerous T-54s (the basis of the T-59 design) were supplied to China by the Soviet Union, this solid MBT will soldier on for many years to come. Some 6,000 examples are still in PLA service, and at least two up-gunned (105mm) versions have been identified by Western military analysts.

Type 69 Main Battle Tank

Country of origin: China.
Crew: 4.
Armament: (Type 69-I) One 100mm smoothbore gun; one Type 54 12.7mm co-axial machine-gun; one Type 59T 7.62mm anti-aircraft machine-gun. **(Type 69-II.)** One 100mm rifled gun; one Type 54 12.7mm co-axial machine-gun; one Type 59T 7.62mm anti-aircraft machine-gun.
Armour: 1.5in-8in (39mm-203mm).
Dimensions: Length (including main armament) 28ft 5in (8.66m); length (hull) 20ft 6in (6.24m); width 10ft 10in (3.29m); height 9ft 3in (2.81m).
Weight: Combat 81,500lb (37,000kg).
Ground pressure: 11.80lb/in² (0.83g/cm²).
Engine: Model 12150L-7BW V-12 liquid-cooled diesel developing 580hp at 2,000rpm.
Performance: Road speed 31mph (50km/h); range 273 miles (440km); vertical obstacle 2ft 7in (0.79m); trench 8ft 10in (2.7m), gradient 60 per cent.
History: First production Type 69 completed in early 1980s. Production continuing. In service with: China (6,000), Iran (several hundred), Iraq (about one thousand), Thailand (several hundred). Production has also recently started in Pakistan.

The Type 69 is a development of the Chinese Type 59, from which it differs mainly in armament, fire control and night vision devices. As in the Type 59, the driver

sits in the front on the left, while the remaining three crewmen are housed in the turret.

The main difference between the Type 69-I and the Type 69-II is in the armament. In the first production batch two different types of gun were installed: some had a 100mm rifled tube and the others a 100mm smoothbore. The tanks with the smoothbore weapon were designated Type 69-I; this had a slightly longer tube, with a bore evacuator near the muzzle. After extensive testing, however, it was concluded that the rifled gun was superior and production of the Type 69-I terminated with the 150th tank.

The Type 69-II has the same rifled 100mm gun as the Type 59 tank, and fires a range of Chinese-designed and produced ammunition, including three types of APFSDS, HEAT, HE and APHE. The gun is fully stabilized, there is a Tank Laser Rangefinder-1 laser mounted above the mantlet and the Tank Simplified Fire Control System (TSFCS) is fitted. Most Type 69-IIs have an externally-mounted laser rangefinder, but the TSFCS-L system incorporates a laser integrated with the gun sight and mounted inside the turret.

The Type 69 has been exported in respectable quantities, Iraq having received several hundred in the early 1980s, while Iran has taken delivery of over 1,000 examples. The Royal Thai Army has ordered some 500 Type 69-IIs under the local designation Type 30.

Below: First seen in 1982, the Chinese Type 69 Main Battle Tank is a development of the Type 59 MBT. Mounted above and to the right of the 100mm gun is a large infra-red searchlight.

AMX-13 Light Tank

Country of origin: France.
Crew: 3.
Armament: One 90mm gun; one 7.5mm or 7.62mm machine-gun co-axial with main armament; two smoke dischargers on each side of turret.
Armour: 0.4in-1.6in (10mm-40mm).
Dimensions: Length (gun forward) 20ft 10in (6.36m); length (hull) 15ft (4.88m); width 8ft 2in (2.50m); height 7ft 7in (2.30m).
Weight: Combat 33,069lb (15,000kg).
Ground pressure: 10.81lb/in² (0.76kg/cm²).
Engine: SOFAM Model 8 GXb eight-cylinder water-cooled petrol engine developing 250hp at 3,200rpm.
Performance: Road speed 37mph (60km/h); range 218 miles (350km); vertical obstacle 2ft 2in (0.65m); trench 5ft 3in (1.6m); gradient 60 per cent.
History: In service with French Army since 1953. Currently in service with following armies: Argentina (80), Chile (47), Dominican Republic (2), Ecuador (108), El Salvador (12), Indonesia (100 +), Ivory Coast (5), Lebanon (50), Morocco (60), Nepal (5), Peru (110), Singapore (350), Tunisia (40) and Venezuela (36). Production complete.

The AMX-13 was designed by the *Atelier de Construction d'Issy-les-Moulineaux* (AMX) near Paris and the first prototype was completed in 1948. The type entered production at the *Atelier de Construction Roanne* in 1952 and production continued at that plant until the early 1960s when it was transferred to the civilian

Below: This AMX-13 Light Tank sports a 105mm gun which can fire a HEAT projectile that can penetrate 14in (360mm) of armour at an incidence of 0° and a range of 3,280ft (1,000m).

Above: This Armoured Recover Vehicle derivative of the AMX-13 Light Tank is used to recover disabled tanks via three winches.

Creusot-Loire plant at Chalons-sur-Saône. The AMX-13 remained in production until the late 1980s by which time no less than 7,700 light tanks, self-propelled guns and armoured personnel carriers (APCs) in this family of armoured fighting vehicles (AFVs) had been constructed. The AMX-13 was designed for use as a tank destroyer and reconnaissance vehicle, and was the standard light tank of the French Army for many years.

The hull is of all-welded steel construction, with a maximum thickness of 1.58in (40mm). The driver sits at the front of the hull on the left, with the engine to his right. The turret is at the rear of the hull, with the commander on the left and the gunner to his right. To keep the hull as low as possible the tank was designed for crew members no taller than 5ft 8in (1.73m). The turret is of an unusual ''oscillating'' design and consists of two parts, the lower one being mounted on the turret ring and has two trunnions. The gun is mounted rigidly in the upper part, which elevates as a complete unit. This design enabled the French to fit an automatic loader and thus reduce the crew from the standard four to three.

The gun is fed from two revolving, 6-round magazines and once 12 rounds have been expended a crew member must physically leave the tank to reload. The empty cartridge cases are ejected through a hatch in the rear of the turret. The first AMX-13s were armed with a 75mm gun firing HE or HEAT rounds. Later models are armed with either a 105mm or a 90mm gun. Many AMX-13s have been fitted with anti-tank missiles, usually the French SS-11. There is a co-axial machine-gun of either 7.5mm or 7.62mm-calibre and there is a mounting on the turret roof for an optional 7.5mm or 7.62mm anti-aircraft machine-gun.

The AMX-13 chassis has been used as the basis for a large number of variants. These include the AMX VCI APC and two self-propelled howitzers: the Mk 61 105mm and Mk F3 155mm. There is also a bridgelayer, the *Char Poseur du Pont*, carrying a Class 25 scissors bridge, and an armoured recovery vehicle, the *Char de Depannaqe*.

The AMX-13 is no longer in use with its two former, largest users, the French and Dutch armies. The Singapore Army is now the largest single operator and Singapore Automotive Engineering is converting these to the new AMX-13 SM1 standard which includes a completely new automative package, with a diesel engine and a fully automatic transmission. An alternative retrofit kit is available from Creusot-Loire, which includes a GIAT 105mm main gun mounted in a Fives Cail Babcock FL-15 turret.

Overleaf: An AMX-13 VCG Engineer Combat Vehicle puts its large, hydraulically-powered dozer blade to good use on rough terrain.

AMX-30B2 Main Battle Tank

Country of origin: France.
Crew: 3.
Armament: One 105mm gun; one 20mm cannon or one 12.7mm machine-gun co-axial with the main armament (see text); one 7.62mm machine-gun on commander's cupola; two smoke dischargers.
Armour: 3.1in (79mm) maximum.
Dimensions: Length (including main armament) 31ft 1in (9.48m); length (hull) 21ft 8in (6.59m); width 10ft 2in (3.1m); height (including searchlight) 9ft 4in (2.85m).
Weight: Combat 79,366lb (36,000kg).
Ground pressure: 12.08lb/in² (0.85kg/cm²).
Engine: Hispano-Suiza HS-110 12-cylinder water-cooled multi-fuel engine developing 720hp at 2,600rpm.
Performance: Speed 40mph (65km/h); range 373 miles (600km); vertical obstacle 3ft 1in (0.93m); trench 9ft 6in (2.9m); gradient 60 per cent.
History: Entered service with French Army in 1967. Now in service with: Chile (21 AMX-30), Cyprus (50 AMX-30B2), France (1,084 AMX-30, most being upgraded to AMX-30B2, plus 271 new-build AMX-30Bs), Greece (190 AMX-30), Qatar (24 AMX-30S), Saudi Arabia (290 AMX-30S), Spain (335 AMX-30), United Arab Emirates (64 AMX-30) and Venezuela (81 AMX-30).

After the end of World War II France quickly developed three vehicles, the AMX-13 light tank, the Panhard EBR 8x8 heavy armoured car and the AMX-50 heavy tank. The last was a very interesting vehicle with a hull and suspension very similar to the German PzKpfw V Panther tank used in some numbers by the French Army in the immediate post-war period. The AMX-50 had an oscillating

turret, a feature that was adopted for the AMX-13 tank. The first AMX-50s had a 90mm gun, this being followed by a 100mm and finally a 120mm weapon. At one time it was intended to place the AMX-50 in production, but as large numbers of American M47s were available under the US Military Aid Program (MAP) the whole programme was cancelled. In 1956 France, Germany and Italy drew up their requirements for a new MBT for the 1960s. The basic idea was good: the French and Germans were each to design a tank to the same general specifications; these would then be evaluated together; and the best tank would then enter production in both countries, for use in all three. But like many international tank programmes which were to follow, this came to nothing: France placed her AMX-30 in production and Germany placed her Leopard 1 in production. The AMX-30 is built at the *Atelier de Construction* at Roanne, which is a government establishment and the only major tank plant in France. The first production AMX-30s were completed in 1966 and entered service with the French Army the following year, where it replaced the American M47. The hull of the AMX-30 is of cast and welded construction, whilst the turret is cast in one piece. The driver is seated at the front of the hull on the left, with the other three crew members in the turret. The commander and gunner are on the right of the turret with the loader on the left. The engine and transmission are at the rear of the hull, and can be removed as a complete unit in under an hour. Suspension is of the torsion-bar type and consists of five road wheels, with the drive sprocket at the rear and the idler at the front, and there are five track-return rollers. These support the inner part of the track. The main armament of the AMX-30 is a 105mm gun of French design and manufacture, with an elevation of +20° and a ▶

Below: The AMX-30 is the standard MBT of the French Army and is manufactured at the *Atelier de Construction Roanne* where the AMX-10P MICV and AMX-10RC recce vehicles were also built.

Below: The AMX-30S was developed specifically for desert operations and has been ordered by Saudi Arabia (290) and Qatar (24). Modifications include the additions of a laser rangefinder, sand shields and a modified engine, as well as a modified transmission which reduces the tank's speed to 37.3mph (60km/h).

▶ depression of −8°, and a traverse of 360°, both elevation and traverse being powered. A 12.7mm machine-gun or a 20mm cannon is mounted to the left of the main armament. This installation is unusual in that it can be elevated independently of the main armament to a maximum of 40°, enabling it to be used against slow-flying aircraft and helicopters. There is a 7.62mm machine-gun mounted on the commander's cupola and this can be aimed and fired from within the turret. Two smoke dischargers are mounted each side of the turret. 47 rounds of 105mm, 500 rounds of 20mm and 2,050 rounds of 7.62mm ammunition are carried. There are five types of ammunition available for the 105mm gun: APFSDS, HEAT, HE, Smoke, Illuminating and Practice. The HEAT round weighs 48.5lb (22kg) complete, has a muzzle velocity of 3,281ft/s (1,000m/s) and will penetrate 14.17in (360mm) of armour at an angle of 0°. Other HEAT projectiles spin rapidly in flight as they are fired from a rifled tank gun, but the French HEAT round has its shaped charge mounted in ball bearings, so as the outer body of the projectile spins rapidly, the charge itself rotates much more slowly. In 1980 an APFSDS projectile entered production, this being able to penetrate 1.96in (50mm) of armour at an incidence of 60° and a range of 5,470 yards (5,000m). The AMX-30 can ford streams to a maximum depth of 6ft 7in (2m) without preparation. A schnorkel can be fitted over the loader's hatch, and this enables the AMX-30 to ford to a depth of 13ft 2in (4m). Infra-red driving equipment is fitted, as is an infra-red searchlight on the commander's cupola and another such searchlight to the left of the main armament. An NBC system is fitted as standard equipment. The latest model of the AMX-30 to enter production for the French Army is the AMX-30B2 which has a number of modifications including a much-improved fire control system. For export the AMX-30 can be delivered without NBC or night-vision equipment and with a much simpler cupola. A special model has been developed for use by Saudi Arabia, this being known as the AMX-30S. It has a laser rangefinder, sand shields

Above: To the left of the AMX-30's main armament is a Sopelem PH-8-B searchlight with a range of 2,623ft (800m) in infra-red mode.

and a modified transmission. There are a number of experimental models of the AMX-30 type and the following models are in service. The AMX-30D is the armoured recovery vehicle and has a crew of four (commander, driver and two mechanics). Equipment fitted includes a dozer blade at the front of the hull, a crane (hydraulically operated) and two winches, one with a capacity of 77,049lb (35,000kg) and the other with a capacity of 8,811lb (4,000kg). Armament consists of a cupola-mounted 7.62mm machine-gun and smoke dischargers. The bridgelayer version carries a scissor-type bridge which, opened out, can span

Above: An example of the versatility of the AMX-30 MBT is this much-modified example used as a transporter/launcher for the Pluton tactical nuclear missile system.

a gap of up to 65ft 7in (20m); this model has a crew of three (commander, bridge operator and driver). The AMX-30 has also been modified to carry and launch the French-developed Pluton tactical nuclear missiles. The missile is elevated for launching and has a maximum range of 62 miles (100km). This model is now in service with the French Army where it replaced the American-supplied Honest John missiles. An anti-aircraft gun tank was also produced for Saudi Arabia, armed with twin 30mm cannon and fitted with an all-weather fire-control system. This has not been adopted by the French Army as it already uses the AMX-13 anti-aircraft tank with a similar turret. There is a separate entry for the 155mm GCT self-propelled gun. Saudi Arabia has also ordered an anti-aircraft missile system called the Shahine, a development of the Crotale missile system which is now in service with the French Air Force and a number of other armies. One AMX-30 vehicle carries six missiles in the ready-to-launch position, as well as the launch radar, whilst another tank has the search and and surveillance radar. The French Army has modified the AMX-30 to carry the Roland SAM system: two missiles are carried in the ready-to-launch position with a further eight missiles inside the hull. Roland has been developed by both France and Germany with the former responsible for the clear weather Roland 1 and the latter responsible for the all weather Roland 2. The AMX-30 has sold reasonably well abroad, but has not been as successful as was once hoped. Later developments aimed solely at the export market were the AMX-32, an upgraded AMX-30 with the same 105mm gun, and the AMX-40, with a 120mm gun. However, neither of these attracted any orders and all French MBT development is now concentrated on the Leclerc.

In a move typical of Western military retrenchment following the end of the Cold War, the French government in August 1991 cancelled plans to upgrade 90 AMX-30s to AMX-30B2 standard. The tanks affected will continue in service unmodified, until replaced by Leclerc MBTs later in the 1990s.

155mm Mk F3
Self-propelled Howitzer

Country of origin: France.
Crew: 2 (+ 8 with accompanying vehicle).
Armament: One 155mm howitzer.
Armour: 0.4in-0.8in (10mm-20mm).
Dimensions: Length (with armament) 20ft 5in (6.22m); width 8ft 11in (2.72m); height 8ft 10in (2.70m).
Weight: (combat) 38,367lb (17,400kg).
Ground pressure: 11.38lb/in² (0.80kg/cm²).
Engine: SOFAM 8-cylinder petrol developing 250hp at 3,200rpm.
Performance: Road speed 37mph (60km/h); range 186 miles (300km); vertical obstacle 2ft 0in (0.6m); trench 4ft 11in (1.5m); gradient 40 per cent.
History: Entered service with the French Army in 1960s. In service with Argentina (24), Chile (12), Ecuador (12), France (222), Kuwait (80), Morocco (64), Qatar (8), Sudan (6), United Arab Emirates (20), Venezuela (20).

The 155mm F3 Self-propelled Howitzer was developed by the *Atelier de Construction de Tarbes* (armament) and the *Atelier de Construction Roanne* (chassis). Production was undertaken by Creusot-Loire.

The chassis is of all-welded steel construction with the driver seated towards

Above: The 155mm self-propelled howitzer Mk F3 is essentially a shortened AMX-13 light tank chassis with a 155mm howitzer mounted at the rear of the hull. It is now being replaced in the French Army by the 155mm GCT.

Right: 155mm self-propelled howitzer Mk F3 with weapon at maximum elevation. The driver and commander ride on the vehicle and the remainder of the gun crew and ammunition follow in an AMX VCA tracked vehicle.

the front on the left and the vehicle commander to his rear; the engine is to the right of the driver and the 155mm howitzer is mounted at the rear. The suspension is of the well-tried torsion bar type and consists of five rubber tyred road wheels with the drive sprocket at the front and the last road wheel acting as the idler, and two track return rollers. The first and last road wheel stations have a hydraulic shock absorber.

The F3 is basically a shortened AMX-13 tank chassis with the 155mm howitzer mounted at the rear. When travelling, the howitzer is in the horizontal position and locked 8° to the right of the vehicle's centreline. The weapon has a double baffle muzzle brake and can be elevated from 0° to +67°; traverse is 20° left and 30°right (with an elevation of 0° to +50°) and 16° left and 30° right (with an elevation from +50° to +67°).

Two of the crew, the driver and commander, ride with the self-propelled howitzer, with the other eight members of the gun crew following in an AMX VCA tracked vehicle which carries 25 projectiles, 25 charges and fuses. The howitzer has a maximum rate of fire of three rounds per minute, although when being used in the sustained fire role this drops to about one round per minute. The ammunition is of the separate loading type, i.e. projectile and charge, and the following types can be fired: HE projectile weighing 96.46lbs (43.75kg), range 21,880 yards (20,000m); hollow base projectile weighing 95.36lbs (43.25kg), range 23,630 yards (21,600m); illuminating projectile weighing 97.02lbs (44kg), range 19,418 yards (17,750m); smoke projectile weighing 97.57lbs (44.25kg), range 19,418 yards (17,750m); and a rocket-assisted projectile weighing 93.71lbs (42.5kg), range 25,162 yards (23,300m).

155mm AU F1 (GCT) Self-propelled Gun

Country of origin: France.
Crew: 4.
Armament: One 155mm gun; one 7.62mm anti-aircraft machine-gun; four smoke dischargers.
Armour: 1.96in (50mm) maximum, estimated.
Dimensions: Length (with gun forward) 33ft 2in (10.2m); length (hull) 21ft 11in (6.7m); width 10ft 4in (3.15m); height (without anti-aircraft machine-gun) 10ft 10in (3.3m).
Weight: 95,899lb (43,500kg).
Ground pressure: 12.8lb/in² (0.9kg/cm²).
Engines: Hispano-Suiza HS-110 12-cylinder multi-fuel engine developing 720hp at 2,400rpm.
Performance: Road speed 37mph (60km/h); range 280 miles (450km); vertical obstacle 3ft 3in (0.93m); trench 6ft 3in (1.9m); gradient 60 per cent.
History: Entered service with the Saudi Arabian Army in 1978, French Army in 1979. In service with France (195), Iraq (85), Saudi Arabia (51).

In the late 1960s the French Army started to look for a successor to the 105mm and 155mm open-mount self-propelled artillery weapons then in service. The four main requirements were laid down as: mobility similar to that of the AMX-30 main battle tank (MBT); ability to engage targets rapidly through a full 360° at all ranges; a high rate of fire with effective ammunition; and protection for the crew from small arms fire an NBC. The first prototype of the new *Grande Cadence de Tir* (GCT) was completed in 1972 and the vehicle entered production in 1977, although an order from Saudi Arabia was met first, with 51 examples being delivered between 1978 and 1982. Deliveries to the French Army started in 1981 and were completed in 1989; 195 are now in service in artillery regiments supporting armoured divisions, each regiment being supplied with four batteries, each of five GCTs. 85 GCTs were delivered to the Iraqi Army in the mid-1980s.

The chassis of the GCT is basically the same as that of the AMX-30 MBT, but with the tank's ammunition racks removed and a new generator and ventilator installed. The tracks and suspension are also the same as on the MBT, as is the Hispano-Suiza HS 110 12-cylinder multi-fuel engine, which can run on petrol, diesel, or paraffin.

The large turret is mounted centrally on the hull and houses three men (the fourth crew member, the driver, sits forward under the glacis plate). The commander and gunner are on the right, while the loader is on the left. The gunner operates the fire control system and sets the elevation and traverse of the gun, while the loader prepares the charges and monitors the automatic loader. The 40-calibre 155mm ordnance was designed by GIAT and is fitted with a large double baffle muzzle brake. Elevation limits are +60° and −4°, while traverse is a full 360°.

The ordnance fires a wide variety of rounds, including all NATO standardised 155m ammunition. The French Army has also developed a range of ammunition specially for this weapon. Among these are conventional high-explosive (HE), base-bleed (extended-range) HE, rocket-assisted projectile (RAP), smoke, illuminating and training shells. The effect of the technological advances in shell design is evident. The standard HE shell delivers a 19.6-lb (8.9kg) payload of HE to a range of 23,184 yards (21,200m). The base-bleed shell, which incorporates special design features to reduce drag, can deliver a slightly greater payload of 22lb (10kg) of HE to the greater range of 31,167 yards (28,500m). The RAP, however, delivers the same 22-lb (10kg) payload out to some 34,450 yards (31,500m), just short of 20 miles (32.18km).

42 projectiles and their separate, bagged charges are carried in the rear of the turret, a typical load consisting of 36 HE and six smoke rounds. The automatic loader enables the GCT to fire at a "burst-rate" of eight rounds per minute.

A machine-gun mounting is located on the turret roof above the loader's hatch and can be used for either a 12.7mm or a 7.62mm anti-aircraft weapon. Two smoke dischargers are mounted on the forward face of the turret. There is also full NBC protection for all the crew.

The tactical concept envisages the gun moving into position, which takes less than two minutes, firing a six-round burst and then coming out of action, again taking less than two minutes, and moving to a new fire position. Such rapidity is essential if the gun is to avoid being subejcted to enemy counter-battery fire.

The GCT is said by some to be too expensive and too heavy. Its combat weight of 95,815lbs (43,500kg) is certainly considerably more than that of the United States' M109A2, which weighs 54,952lbs (24,948kg). However, it is popular in the French Army, which procured 195 at a fairly slow rate between 1981-89.

Left: 155mm GCT SPG from the rear with its ammunition resupply hatches lowered to show ammunition racks for 42 projectiles (eg., 36 HE and 6 smoke) and 42 charges. Four men can reload the GCT within 20 minutes.

Below: 155m GCT SPG with turret traversed to the rear. It is the first SPG in NATO to have an automatic loading system which enables it to fire eight projectiles per minute to a maximum range of 34,450 yards (31,500m).

Below: A quartet of French Army GCTs raise their 155mm guns and commence laying down a barrage of fire. A maximum of 42 projectiles, along with their bagged charges, can be carried; maximum rate of fire is eight rounds per minute.

Leclerc Main Battle Tank

Country of origin: France.
Crew: 3.
Armament: One GIAT CN 120-26 120mm smoothbore gun; one 12.7mm coaxial machine-gun; one 7.62mm anti-aircraft machine-gun; nine smoke grenade dischargers.
Armour: Spaced, multi-layer.
Dimensions: Length (including main armament) 32ft 5in (9.87m); length (hull) 22ft 7in (6.88m); width 12ft 2in (3.71m); height 8ft 1in (2.46m).
Weight: Combat 116,700lb (53,000kg).
Ground pressure: 12.80lb/in² (0.90kg/cm²).
Engine: Uni-Diesel V8X-1500 Hyperbar 8-cylinder liquid-cooled diesel engine developing 1,500hp; Turbomeca gas turbine auxiliary power unit.
Performance: Road speed 44mph (71km/h); range 341 miles (550km); vertical obstacle 4ft 1in (1.25m); trench 9ft 10in (3m), gradient 60 per cent.
History: Entered production in 1990; will enter service with French Army in 1992.

In the late 1970s the French and West German armies undertook a second collaborative project to develop a replacement for the AMX-30 and Leopard 1 MBTs, which were themselves the separate outcomes of a previous attempt at a joint design. But, like all previous collaborative tank ventures, this one also collapsed (in December 1982) and the French Army set about the design of a new tank for the 1990s designated *Engin Principal de Combat (EPC)*. Project definition was completed in 1986 and the first prototype was up and running by the end of 1989. By this time the tank had been named the Leclerc after one of the most successful of French World War Two commanders.

The Leclerc is generally similar in size and armament to other contemporary Western MBTs, except that, unlike the Leopard 2 (Germany), Challenger 2 (UK) and M1 (USA), it has a crew of only three men. This is achieved through the replacement of the fourth crewman by an automatic loader, a feature the Leclerc shares with the Japanese Type 90 and the Soviet T-64/-72/-80.

The hull and turret shells are constructed of welded steel to which is added modular segments of composite armour. The current armour is claimed to have high resistance to kinetic energy and chemical rounds, but being modular can be replaced by new or improved armour later in the tank's life, if required. The original turret fitted to the prototypes was very angular in shape, but the production tanks sport a very long and low turret of excellent ballistic shape, which offers armour protection for the laser, machine-gun and even the grenade launchers.

The driver sits well to the rear of the glacis plate and just to the left of the centreline of the tank, with a drum of 18 rounds of 120mm ammunition to his

Above: The shape of the French Army's Main Battle Tank of the 1990s and beyond, the Leclerc, with a 120mm smoothbore gun. Procurement plans call for 850 Leclercs to be built.

Above: A prototype Leclerc MBT being put through its paces. Of particular note is the early turret — quite different in shape to that sported by the Leclerc on the previous page.

right. The commander is in the turret to the left of the main gun, with the gunner on the right, the reverse of the arrangement in most other MBTs where the commander is on the right. The crew is supplied with a sophisticated, computerised battle management system, which controls and monitors all activities within the vehicle, and which will even give status reports to a higher headquarters either at given intervals or on request.

The gun is a 120mm smoothbore, designed and built by GIAT. Its chamber is the same size as that on the German and US 120mm guns, thus ensuring commonality of ammunition, but the tube is somewhat longer (52 as opposed to 44 calibres), which imparts a higher muzzle velocity to the projectiles, particularly APFSDS. The gun has a thermal sleeve, but rather than a fume extractor fitted to the barrel it uses a compressed-air system to expel fumes automatically after firing. The main guns on the prototype vehicles were fitted with a muzzle reference system, but this does not appear to be fitted to the production models.

The automatic loader is made by Creusot-Loire and contains 22 rounds of ready-use ammunition. The system can distinguish between five different natures of ammunition and will load the appropriate round as selected by the commander or gunner. To reload, the gun automatically returns to the -1.8° position and then resumes the elevation directed by the commander or gunner. The system is claimed to be capable of a firing rate of 15 rounds per minute, although the normal maximum would be about 12. The automatic loader is mounted in the long turret bustle and is separated from the fighting compartment by a bulkhead, and there are blow-out panels in the roof to divert any explosion away from the crew.

There are currently two principle rounds. The APFSDS round has a tungsten penetrator with a muzzle velocity (mv) of 5,742ft/sec (1,750m/sec), while the HEAT round has an mv of 3,610ft/sec (1,100m/s).

The French mounted a 20mm coaxial cannon in the AMX-30/-32 and AMX-40 series of MBTs, which could elevate independently of the main armament. In the Leclerc they have dropped the idea of independent elevation, but the coaxial weapon is a 12.7mm machine-gun, which is again unusual, since virtually all other MBTs use 7.62mm calibre machine-guns for this application. A 7.62mm machine-gun is located on the turret roof, which has an armoured casing and is fully-controlled from within the tank.

The power unit is a 1,500hp Uni-Diesel diesel engine, which gives the very high power:weight ratio of 28.3hp/tonne. There is also a Turbomeca gas turbine, which is used to provide power when the tank is stationary, so that the main engine can be closed down.

Jagdpanzer 4-5 Kanone, Jagdpanzer Jaguar 1 Rakete and Jagdpanzer Jaguar 2

Country of origin: Germany.
Armament: Kanone One 90mm rifled gun; one MG3 7.62mm co-axial machine-gun; one MG3 7.62mm anti-aircraft machine-gun. **Jaguar 1** Two Aerospatiale SS-11 ATGW launchers or one Euromissile K3S ATGW launcher; one bow-mounted MG3 7.62mm machine-gun; one MG3 7.62mm anti-aircraft machine-gun. **Jaguar 2** One Hughes TOW ATGW launcher; one MG3 7.62mm anti-aircraft machine-gun.
Armour: 0.5in-2in (12.7mm-51mm).
Dimensions: Length (including barrel) 28ft 9in (8.75m); length (hull) 20ft 6in (6.24m); width 9ft 9in (2.98m); height 6ft 10in (2.09m).
Weight: Combat 60,573lb (27,500kg).
Engine: One Daimler-Benz Model MB 837 eight-cylinder water-cooled diesel developing 500hp at 2,000rpm.
Performance: Road speed 44mph (70km/h); range 250 miles (400km); vertical obstacle 2ft 6in (0.75m); trench 6ft 7in (2.0m); gradient 60 per cent.
History: Kanone entered service with the German Army (750) in 1965 and with the Belgian Army (80) in 1975. **Jaguar 1** entered service with the German Army (370) in 1967. **Jaguar 2** entered service with the German Army (162) in 1983.

During World War II the German Army (*Wehrmacht*) had considerable success with their Jagdpanzer (Jpz: "tank destroyers") and when the force was reformed in 1954 it ordered development of a new type, to be armed with a 90mm cannon and designated Jpz 4-5. After a slow development process this new vehicle entered production in 1965, with 750 being produced. The Jpz 4-5 uses a chassis which was shared with the Jpz Rakete (see below), the Marder infantry combat vehicle (ICV) and an abortive reconnaissance tank.

Below: This Jagdpanzer Jaguar 1 Rakete clearly displays its original Aerospatiale SS-11 ATGW weaponry, as well as a very prominent sighting system. The vast majority of the Jaguar 1s were subsequently rearmed with the Euromissile HOT ATGW.

The Jpz 4-5 has a very low hull, with the 90mm cannon mounted in the glacis plate, where it is slightly offset to the right. The gun controls are manual, with traverse 15° left or right of centre and elevation limits of + 15°/ – 8°. It fires HEAT, HESH-T or HEAT-P rounds. There are eight electrically-operated smoke grenade launchers, which are mounted on the engine-decking in a fan-shape. The four-man crew comprises the commander, driver, loader and gunner.

A virtually identical hull is used for the Jpz Rakete, which is also known as the Jaguar 1. 370 were built between 1967-68, all armed with two Aerospataile SS-11 missile launchers. Some years later, following successful trials, 316 of the 370 Jaguar 1s were converted to take the Euromissile HOT missile, which has a higher speed, greater range and an improved rate of fire; it is also simpler to operate, since all the gunner has to do is to keep the sight on the target. Fewer missiles can be carried, however: nine HOT missiles compared to 14 SS-11s. The improved HOT 2 replaced the HOT 1 in 1985. The HOT-armed Jaguar 1 also has additional spaced armour covering the front and sides of the fighting ▶

Below: The Jagdpanzer 4-5 Kanone is used by the armies of Germany and Belgium, and can be readily identified by virtue of its 90mm gun. 51 rounds of ammunition are carried for the gun.

compartment. A 7.62mm MG3 machine-gun is mounted in the right-hand side of the glacis plate.

The Jaguar 2 is similar to Jaguar 1, but has the Hughes TOW missile in place of the HOT. The very successful and widely used TOW missile is considerably cheaper than the HOT; it has a similar range to the HOT, but its greater speed reduces the time of flight. In 1983-85 the German Army converted 162 of the 750 Jpz Kanone to Jaguar 2 configuration, by removing the gun and covering the glacis with a large plate of spaced armour. No bow machine-gun is fitted in this version.

1,120 Jpzs were built for the German Army. Today there are 542 Jpz Rakete vehicles: 162 Jaguar 2 with TOW, 316 of the improved Jaguar 1 with HOT and 64 of the original Jaguar 1 with SS-11. There are also some 578 Jpz Kanone remaining, although some of these have been converted to forward observation vehicles by removing the 90mm gun.

The Belgian Army ordered a development of the Jpz 4-5, which was produced

Below: The Jagdpanzer 4-5 Kanone's 90mm rifled gun is set in the glacis plate and has a traverse of 15° left or right of centre and can be elevated from − 8° to + 15°; both traverse and elevation are conducted manually. Belgian Army Kanones (known as JPK-90s) use a laser rangefinder when firing.

in Belgium. It is very similar to Jpz 4-5, except that it has the suspension and tracks of the Marder ICV. Known as the JPK-90, there are 80 such vehicles in service with the Belgian Army.

Below: A Jagdpanzer Jaguar 1 Rakete armed with a Euromissile HOT ATGW in the ready-to-launch position. Also visible is the MG3 7.62mm anti-aircraft machine-gun atop the vehicle.

Middle and above: A Jagdpanzer 4-5 Kanone being put through its paces in the German countryside. Just visible in the upper shot is the bank of electrically-operated, forward-firing smoke dischargers mounted over the rear of the hull. The step in the hull identifies the divide between engine (rear) and combat (forward) stations.

Leopard 1 Main Battle Tank

Country of origin: Germany.
Crew: 4.
Armament: One 105mm gun; one 7.62mm machine-gun co-axial with main armament; one 7.62mm machine-gun on roof; four smoke dischargers on each side of the turret.
Armour: 0.4in-2.8in (10mm-70mm).
Dimensions: Length (including main armament) 31ft 4in (9.54m); length (hull) 23ft 3in (7.09m); width 10ft 8in (3.25m); height 8ft 8in (2.64m).
Weight: Combat 93,394lb (42,400kg).
Ground pressure: 12.23lb/in² (0.86kg/cm²).
Engine: MTU MB 838 Ca.M500 10-cylinder multi-fuel engine developing 830hp at 2,200rpm.
Performance: Road speed 40mph (65km/h); range 373 miles (600km); vertical obstacle 3ft 9in (1.15m); trench 9ft 10in (3m); gradient 60 per cent.
History: In service with German Army since 1967. Also ordered (MBT only) by: Australia (90), Belgium (334), Canada (114), Denmark (120), Germany (2,437), Greece (106), Italy (920), the Netherlands (468), Norway (78) and Turkey (77).

The German Leopard 1 MBT has been one of the most successful Western tanks of the post-war era, with some 4,800 MBTs and 1,772 other versions being ordered by 10 armies. In fact, at one point the Leopard 1 appeared to be poised to become the *de facto* NATO standard tank, being used by nine of the 15 NATO armies.

When it was reformed in the 1950s, the West German Army (*Bundeswehr*) was equipped with American M47 MBTs and it was proposed to replace these with a Franco-German collaborative project. Unfortunately, as has happened with almost every MBT collaborative project, there were disagreements and the partners split up, the French developing the AMX-30 and the Germans the Leopard 1. In 1963 the production contract was placed with Krauss-Maffei of Munich and the first production Leopard 1 was delivered in September 1965. Krupp-Mak of Kiel also produced some Leopard 1 MBTs and most of the armoured

Below: Leopard 1A3 with infra-red/white searchlight mounted over the 105mm gun that was developed for the Centurion tank.

Above: Early production Leopard 1 MBT of the German Army being recovered by a Leopard armoured recovery vehicle built by MaK.

engineer vehicles, bridgelayers and recovery vehicles. The production lines closed in late 1979, but were reopened in 1981 to meet new orders from Greece and Turkey. A third production line was set up by OTO Melara in Italy to meet most of the Italian Army's order.

The Leopard 1 has a crew of four, with the driver in the front of the hull and to the right, and the other three located in the turret. The main armament of all versions of the Leopard 1 is the British L7A3 gun manufactured by Royal Ordnance at their factory in Nottingham, England. 55 rounds of 105mm ammunition are carried; 13 in the turret and 42 in the hull. There are two Rheinmetall MG3 7.62mm machine-guns, one mounted co-axially with the main armament and the second in a flexible mounting on the turret roof. ▶

Below: The Leopard 1A3 has a number of improvements including a new all-welded steel turret which gives increased protection.

Above: A dramatic study of a Leopard 1 MBT at full speed. It can reach a top speed of 40mph (65km/h) on made-up roads, a little slower than the Leopard 2 MBT.

► Standard equipment includes night vision devices, an NBC system and a crew heater. The vehicle can ford to a maximum depth of 7ft 5in (2.25m) with minimum preparation using a short shaft fitted over the commander's hatch, and to 13ft 2in (4m) with a longer shaft. Such equipment is, however, very rarely used and, except in the most unusual circumstances, rivers are crossed by bridge of ferry.

The first 1,845 Leopard 1s were built for the *Bundeswehr* in four production batches. All of these were subsequently modified with the installation of a thermal sleeve on the main armament, a gun stablisisation system, new tracks and skirts, becoming the Leopard 1A1 in the process. All of these were then further modified with appliqué armour on the turret and gun mantlet to become what is known as the Leopard 1A1A1.

The fifth production batch consisted of 342 tanks. 232 of these were built to the Leopard 1A1 standard with, in addition, a stronger turret, better NBC system and passive image intensification (II) equipment for the commander and driver; these were designated Leopard 1A2. The other 110 tanks of the fifth batch were

built with all the improvements of the Leopard 1A1 and 1A2, plus the additional armour of the Leopard 1A1A1; these were designated Leopard 1A3.

The final production batch for the *Bundeswehr* were 250 Leopard 1A4s, which were essentially Leopard 1A3s with an integrated fire control system added. Work is currently in hand on a conversion programme for existing tanks which involves fitting a new fire control system, a thermal imaging device, further appliqué armour and other protective devices. 1,300 of the *Bundeswehr's* fleet are being modified; when the work is completed on each tank it is redesignated Leopard 1A5.

The various types of Leopard 1 will remain in service with the *Bundeswehr* well into the 21st Century, with additional improvements being incorporated as they become necessary. Two Leopard 1s were converted in the late 1980s to take a Rheinmetall 120mm smoothbore gun, but despite satisfactory trials the work has proceeded no further.

The Leopard 1 MBT has proved a major success in overseas markets, as listed. One of the most significant of these was the Australian Army order for 90 Leopard 1A3s under the designation Leopard AS1, which were delivered between 1976 and 1978. This was not just the only order from a Far Eastern customer, but was also the only order to be accepted from a non-NATO country.

Leopard 2 Main Battle Tank

Country of origin: Germany.
Crew: 4.
Armament: One Rheinmetall 120mm smoothbore gun; one MG3 7.62mm co-axial machine-gun; one MG3 7.62mm anti-aircraft machine-gun; 16 smoke dischargers (eight on each side of turret).
Armour: Spaced, multi-layer.
Dimensions: Length (including main armament) 31ft 9in (9.67m); length (hull) 25ft 4in (7.72m); width 12ft 2in (3.7m); height 8ft 2in (2.48m).
Weight: Combat 121,475lb (55,150kg).
Ground pressure: 11.80lb/in² (0.83kg/cm²).
Engine: MTU MB-873 Ka-501 12-cylinder liquid-cooled diesel engine developing 1,500bhp at 2,600rpm.
Performance: Road speed 45mph (72km/h); range 600 miles (550km); vertical obstacle 3ft 7in (1.1m); trench 9ft 10in (3m); gradient 60 per cent.
History: Entered service with German Army in 1980. Ordered by: Germany (1,800), the Netherlands (445) and Switzerland (380).

The development of the Leopard 2 MBT can be traced back to a project started in the 1960s. At this time the Germans and the Americans were still working on the MBT-70 programme, so the project had a very low priority. Once the MBT-70 was cancelled in January 1970, the Germans pushed ahead with the Leopard 2, and 17 prototypes were completed in 1974. These prototypes were built by the manufacturers of the Leopard 1, Krauss-Maffei of Munich, with the assistance of many other German companies. Without doubt, the Leopard 2 is one of the most advanced tanks in the world and the Germans have succeeded in designing a tank with high success in all three areas of tank design: mobility; firepower and armour protection. In the past, most tanks have only been able ▶

Above: A Leopard 2 is put through its paces in a typically muddy European tank training area. In 1977 the German Army ordered 1,800 of these excellent MBTs, the first of which was delivered by MaK of Kiel in late 1979.

Left: Main armament of the Leopard 2 is a 120mm Rhein-Metall smoothbore gun which fires two types of fixed ammunition, APFSDS (Armour-Piercing Fin-Stabilised Discarding Sabot) and HEAT-MP (High-Explosive Anti-Tank Multi-Purpose). A total of 42 rounds of 120mm ammunition are carried.

Above: Protruding from cover, this Leopard 2 presents a small, low target to enemy anti-tank gunners. In addition to being ordered by the German Army, the Dutch and Swiss Armies have ordered no less than 825 Leopard 2s between them.

The Leopard 2 has proved itself capable of high mobility over all types of terrain. Its survivability on the battlefields of the 1990s could depend on its high protection and agility.

▶ to achieve two of these objectives at once. A good example is the British Chieftain, which has an excellent gun and good armour, but poor mobility; the French AMX-30 is at the other end of the scale and has good mobility, an adequate gun but rather thin armour. The layout of the Leopard 2 is conventional, with the driver at the front, turret with commander, gunner and loader in the centre, and the engine and transmission at the rear. The engine was in fact originally developed for the MBT-70. The complete powerpack can be removed in about 15 minutes for repair or replacement. At first it was widely believed that the Leopard 2's armour was of the spaced type, but late in 1976 it was revealed that it used the British-developed Chobham armour. This gives superior protection against attack from all known projectiles. It is of the laminate type, and consists of layers of steel and ceramics. The suspension system is of the torsion-bar type with dampers. It has seven road wheels, with the drive sprocket at the rear and the idler at the front, and there are four track return rollers. The first prototypes were armed with a 105mm gun of the smoothbore type, developed by Rheinmetall, but later prototypes had the 120mm smoothbore gun. The 120mm gun fires two basic types of fin-stabilised ammunition (in which small fins unfold from the rear of the round just after it has left the barrel), and this means that the barrel does not need to be rifled. The anti-tank round is of the Armour-Piercing Discarding Sabot type, and has an effective range of well over 2,405 yards (2,200m); at this range it will penetrate a standard NATO heavy tank target. The second round is also fin-stabilised and is designed for use against field fortifications and other battlefield targets. The cartridge case is semi-combustible and only the cartridge stub, which is made of conventional steel, remains after the round has been fired. The job of the loader is eased by the use of the hydraulically-assisted loading mechanism. The gun has an elevation of + 20° and a depression

of − 9°. A standard 7.62mm MG3 machine-gun is mounted co-axially with the main armament. A 7.62mm MG3 machine-gun is installed on the loader's hatch for use in the anti-aircraft role. 42 rounds of 120mm and 2,000 rounds of 7.62mm ammunition are carried. Eight smoke dischargers are mounted each side of the turret. A very advanced fire-control system is fitted, which includes a combined laser and stereoscopic rangefinder, and the gun is fully stabilised, enabling it to be laid and fired on the move with a high probability of the round hitting the target. Standard equipment includes infra-red and passive night-vision equipment, an NBC system and heaters for both the driver's and fighting compartments. The Leopard 2 can ford streams to a depth of 2ft 7in (0.8m) without preparation, and with the aid of a schnorkel can deep ford to a depth of 13ft 1in (4m).

The Leopard 2 has established an enviable reputation and has been tested by a number of armies. The US Army evaluated a special tank designated "Leopard 2 Austere Version" against the US-designed XM1, from which the later emerged victorious. In the late 1980s the British also examined a different version known as Leopard 2 (Improved), but again this failed to win an order against the national rival, in this case the Challenger 2. The Leopard 2 has also been tested in Sweden, whose army is currently seeking a replacement for the S-Tank.

The German tank has, however, won two foreign contracts. In 1979 the Dutch Army ordered 445 Leopard 2s, which were delivered between 1982 and 1986. The other order was placed by the Swiss. 35 Leopard 2s were delivered from Germany in 1987, while the remaining 345 are being built in Switzerland, with the last due off the production line in 1993.

No major variants of the Leopard 2 have entered production, although a driver training version has been produced, with a special cab in place of the turret. A prototype of an armoured recovery vehicle, designated *Bergepanzer 3*, has been built and tested, but no orders have been placed.

Design work has started on the next German tank, designated *Kampfpanzer 2000*. This is likely to be armed with a 140mm smoothbore gun and to have a three-man crew.

Merkava Main Battle Tank

Country of origin: Israel.
Armament: One M68 105mm rifled gun; one co-axial 7.62mm machine-gun; two 7.62mm anti-aircraft machine-guns; one 60mm mortar.
Armour: See text.
Dimensions: Length (gun forward) 28ft 5in (8.65m); length (hull) 24ft 5in (7.45m); width 12ft 2in (3.70m); height (commander's cupola) 9ft 0in (2.75m).
Weight: Combat 132,160lb (60,000kg).
Engine: Teledyne Continental AVDS-1790-6A V-12 diesel developing 900hp.
Performance: Road speed 29mph (46km/h); range 250 miles (400km); vertical obstacle 3ft 1in (0.95m); trench 9ft 10in (3.0m); gradient 60 per cent.
History: In service with the Israeli Army.

Below: On patrol in southern Lebanon, an Israeli Army Merkava Mark 1 reveals the type's extremely low-profile turret.

Since the State of Israel was established in 1949 its Army has had more experience of armoured warfare than any other in the world. The Israeli Armoured Corps has always been forced by circumstances to use a mixture of tank types, some bought overseas and others captured during one of the numerous Arab-Israeli wars. Thus, in 1991 the foreign tanks in service came from three principle sources: British-supplied Centurions (1,080), US-supplied M48A5s (550) and M60s (1,400), and Soviet tanks captured from Arab armies, comprising T-54/T-55s (488) and T-62s (110).

The Israelis began to develop their own armoured warfare doctrines based on their increasingly extensive combat experiences and quickly integrated these disparate tank designs into their overall scheme. Nevertheless, the aim was to design their own tank, which would be tailored to their own needs. Being a small country with much less of a manpower base than their larger neighbours, the Israelis cannot afford large losses and it became particularly evident to them in the 1967 campaign that they needed to give armoured protection top priority in any future MBT design. This left firepower as the second priority and mobility third. The outcome was the Merkava (*Chariot*), the first prototype of which was ▶

completed in 1974, although it was not revealed publicly until 1977. The first production vehicles were issued to the Army in 1979 and the type saw initial combat in Lebanon in 1982. Following these early experiences a modified vehicle was developed, designated the Mark 2, and all Mark 1s are being gradually brought up to this standard, with some 600 Merkavas of both marks currently in service with the Israeli Army.

The majority of modern MBTs have the engine at the rear, but the layout of the Merkava is unusual, with the engine and transmission at the front of the tank. This is intended to increase the protection for the crew, since the Israeli Army would much rather save the crew and lose the tank. The hull is made of an outer layer of cast armour, with a welded inside layer, the space between the two being filled by diesel fuel. The driver is seated forward and on the left, with the engine to his right. The engine is a Teledyne Continental AVDS-1790-6A, a more powerful version of that fitted in the US Army's M60, which is also used by the Israeli Army. The Mark 1 used the standard Allison CD-850-6BX semi-automatic transmission, but this has been replaced in the Mark 2 by an Israeli-designed Ashot system whose efficiency is such, it is claimed, that it results in a substantial increase in range. The suspension and road wheels are similar to those used by the Israeli Army's Centurions and there are six road wheels with the drive sprockets at the front and idlers at the rear. There are return rollers and the tops of the tracks have steel covers backed by plates of "special" armour to protect them and the suspension from damage by HEAT weapons.

The turret has an exceptionally small cross-section and a well-sloped front, presenting an extremely small target when the tank is in a hull-down position. There is a layer of "special" armour on the turret front and sides.

The commander and gunner are seated on the right of the turret, with the loader on the left. The main armament is an Israeli-produced version of the M68 105mm gun, which is itself a licence-produced version of the British L7. The barrel is fitted with a thermal sleeve and a fume extractor. This gun is also mounted in the Israeli Centurion, M48 and M60 MBTs, as well as on most of the captured T-54/-55 and T-62s.

The gun has an elevation of + 20° and a depression of -8.5°. There is a travelling lock for the gun on the right forward engine deck. The gun fires all the standard 105mm projectiles, as well as the specially-developed M111 APFSDS-T projectile developed by Israel Military Industries and the more recent M413 which has a maximum effective range of about 6,500 yards (5,950m). No less than 85 rounds are carried, significantly more than in any other modern MBT.

Above: A view inside the turret of a Merkava Mark 1, with the gunner's station, offset to the right side of the turret, in view. The peritelescope includes a moving mirror head and electro-optical laser-type rangefinder, magnifications ranging from x8 to x1.

Left: Putting the Merkava's suspension through its paces, a pair of tanks advance over rough terrain on the Golan Heights to take up position in a wood. Visible in the vertical hull rear is a most interesting feature — a rear entry/exit hatch, enabling the crew to escape from the tank and also bestowing the tank with a transport role.

There is one co-axial 7.62mm machine-gun and two 7.62mm anti-aircraft machine-guns (AAMGs) on flexible mounts on the turret roof; all are of the MAG type, built under licence from FN in Belgium. It is reported that some Merkavas have been fitted with a remotely-operated 12.7mm machine-gun in place of one of the two 7.62mm AAMGs. A unique feature is a 60mm mortar, which is loaded and fired from within the turret. Capable of firing HE, smoke and illuminating rounds, this is intended to save valuable 105mm ammunition, and, again, is an interesting result of Israeli combat experience.

Placing the engine at the front creates considerable space at the rear of the tank. This is normally used to house ammunition and the Merkava carries a normal load of 62 rounds, although this can be increased to 85, if required to do so. The space can, however, also be used to accommodate infantrymen or commandos, although this is only done in exceptional circumstances. Additional communications facilities can also be installed in place of some of the ammunition to enable the tank to be used as a command post.

In 1989 the Merkava Mk 3 was revealed, which is already in service with the Army. This is armed with a 120mm smoothbore gun, for which 50 rounds are carried. An automatic loader (not yet in service) is being developed for this weapon, which is designed to assist but not to replace the loader. Other improvements include all-electrical controls and a new series of laser rangefinder. The armour is of a modular design which is not only more effective than that on the Marks 1 and 2 but can also be changed in the field. A new suspension is also fitted, which, in combination with the uprated engine and improved transmission, gives much-enhanced cross-country mobility.

Left: A factory-fresh Merkava Mark 3, the latest variant to enter service with the Israeli Army. Note the entry/exit door.

Below: A Merkava Mark 3 in side elevation, revealing the 120mm smoothbore gun adopted in place of the M68 105mm rifled gun.

Soltam L-33 155mm Self-propelled Gun/Howitzer

Country of origin: Israel.
Crew: 8.
Armament: One 155mm gun/howitzer; one 7.62mm anti-aircraft machine-gun.
Armour: 0.4in-2.5in (12mm-64mm).
Dimensions: Length (with armament) 27ft 9in (8.47m); length (hull) 21ft 3in (6.47m); width 11ft 4in (3.45m); height 11ft 4in (3.45m).
Weight: Combat 91,507lb (41,500kg).
Engine: Cummins VT 8-460-B1 diesel developing 460hp at 2,600rpm.
Performance: Road speed 22mph (36km/h); road range 161 miles (260km); vertical obstacle 2ft 11in (0.91m); trench 7ft 6in (2.3m); gradient 60 per cent.
History: Entered service with the Israeli Army in 1970/71. Production complete.

In the late 1960s the Israeli Soltam Company started to develop a new self-propelled gun/howitzer for the Israeli Army. This had to be based on the Sherman chassis, provide all round protection for the crew, have a high rate of fire, a good range and carry an adequate supply of onboard ammunition which could be easily replenished. After trials with prototype weapons the Soltam design was accepted for service with the Israeli Army as the L-33.

The L-33 is essentially a much-modified M4A3E8 Sherman tank chassis with the turret removed, engine moved forwards and a new all welded superstructure added. The driver is seated at the front of the hull on the left and the commander seated above and to his rear, both provided with bulletproof windows. An entry door is provided in each side of the hull and there are two hatches in the roof, one for the commander on the left and one for the anti-aircraft gunner on the right. Mounted at the anti-aircraft gunner's station is a 7.62mm machine-gun with a traverse of 360°.

The 155mm gun/howitzer is mounted in the front of the vehicle and has a maximum elevation of +52°, a depression of −3° and a traverse of 30° left and 30° right, elevation and traverse both being manual. The ordnance, which is based on the M-68 towed gun/howitzer, has a fume extractor, single baffle muzzle brake and a pneumatic rammer which enables the weapon to be loaded at all angles of elevation. It fires an HE projectile weighing 96.35lb (43.7kg) with a maximum muzzle velocity of 740 yards a second (725m/s) to a maximum range of 21,880 yards (20,000m); other types of ammunition that can be fired include smoke, practice and illuminating. A well-trained crew can fire four rounds a minute for a short period and a total of 60 155m projectiles are carried, of which 16 are ready for use. Doors are provided in the rear of the hull to facilitate the rapid resupply of ammunition. The exact number of L-33s acquired by the Israeli Army is unknown, but it is known that the production/conversion programme (carried out by Haifa-based Soltam) is complete. To date, no known variants have been identified or accepted for service with the Israeli Army.

Above: A Soltam L-33 15mm Self-propelled Gun/Howitzer viewed from the rear with the resupply doors in the hull in the open position. A total of 60 155mm projectiles and charges can be carried, of which 16 are ready to use. The weapon fires an HE projectile to a maximum range of 21,880 yards (20,000m).

Left: A Soltam L-33 155m Self-propelled Gun/Howitzer showing the weapon which is mounted in the forward part of the hull with a depression of −3°, an elevation of +52°, and a traverse of 30° left or right. The L-33's combat debut came during the 1973 Yom Kippur War.

C1 Ariete Main Battle Tank

Country of origin: Italy.
Armament: One OTO Melara 120mm smoothbore gun; one 7.62mm co-axial machine-gun; one 7.62mm anti-aircraft machine-gun.
Armour: See text.
Dimensions: Length (gun forward) 31ft 3in (9.52m); length (hull) 24ft 11in (7.60m); width 11ft 10in (3.6m); height (commander's cupola) 8ft 0in (2.45m).
Weight: Combat 105,725lb (48,000kg).
Engine: One Fiat V-12 MTCA turbo-charged inter-cooled 12-cylinder diesel developing 1,200hp.
Performance: Road speed +40mph (+65km/h); range +342 miles (+550km); vertical obstacle 6ft 10in (2.1m); trench 9ft 10in (3.0m); gradient 60 per cent.
History: In production for the Italian Army.

Following the end of World War II the Italian Army used American MBTs for many years. They ordered the M60A1, of which 100 were supplied from US arsenals and a further 200 were locally-built by OTO Melara. The Italians were, however, interested in a European tank and had some contact with France and Germany during the European Tank project which resulted, after the nations had agreed to split, in the AMX-30 Leopard 1. In 1970 the Italian Army ordered Leopard 1s from Germany. The initial order was for 200 Leopard A1 MBTs, which were delivered from Krupp-MaK in 1971/72, followed by 69 ARVs and 12 AEVs. A further 720 MBTs, 68 ARVs, 28 AEVs and 64 AVLBs were built by OTO-Melara under a licence agreement.

Below: The Italian-designed and built C1 Ariete represents a major step forward in the country's indigenous tank-building programme. It will enter full service during the early 1990s.

OTO-Melara, in conjunction with Fiat then designed the first Italian post-war MBT, the OF-40 (O = OTO-Melara, F = Fiat, 40 = approximate combat weight in tons). This neat and powerful-looking tank was an original design, but bore some outward resemblance to the Leopard 1A4 and was designed specifically for export. Total sales amounted to 36 MBTs and 3 ARVs, all for the United Arab Emirates.

The Italian Army issued a requirement for a new MBT in 1982, for which one of the criteria was that it must be manufactured in Italy. Design work started in 1984 and the OTO-Melara/Iveco Fiat consortium had the first prototype up and running in 1986, with the remaining five prototypes completed by 1988.

The C1 has a combat weight of 105,725lb (48,000kg) and is armed with a 120mm smoothbore gun, which has been designed and manufactured in Italy by OTO Melara. Like other such weapons, the tube is fitted with a muzzle reference system and fume extractor, and is covered by a thermal sleeve. The chamber is identical with that on the Rheinmetall 120mm smoothbore, thus ensuring that the same ammunition can be used in both types of weapon. The gun fires APFSDS and HEAT-MP rounds, together with smoke and illuminating rounds, a total of 42 being carried, 15 in the bustle and 27 in the hull. The gun is mounted in a fixed mantlet and has a maximum elevation of 20° and depression of -9°. A two-axis stabilization system is fitted. There are also eight smoke dischargers, four on each side of the turret.

The layout of the four-man crew is conventional. The driver sits under the glacis plate forward and to the right. The remainder of the crew are in the turret with the commander and gunner on the right of the main gun and the loader to its left. The hull is of all-welded steel construction and there is a layer of "advanced armour" on the nose and glacis plate. The turret is well-shaped with a well-angled face, unlike some modern tanks (such as the Leopard 2) which have a vertical face. Turret traverse is hydro-electric, with manual back-up.

The torsion-bar suspension system has seven roadwheels with four return rollers. The suspension and upper part of the track are protected by skirts.

Type 61 Main Battle Tank

Country of origin: Japan.
Crew: 4.
Armament: One 90mm gun; one M1919A4 .3in machine-gun co-axial with main armament; one M2 .5in anti-aircraft machine-gun.
Armour: 2.52in (60mm) maximum.
Dimensions: Length (overall) 26ft 10½in (8.19m); length (hull) 20ft 8in (6.3m); width 9ft 8in (2.95m); height 10ft 4in (3.16m).
Weight: Combat 77,162lbs (35,00kg).
Ground pressure: 13.5lb/in² (0.95kg/cm²).
Engine: Mitsubishi Type 12 HM 21 WT 12-cylinder diesel developing 600hp at 2,100rpm.
Performance: Road speed 28mph (45km/h); range 124 miles (200km); vertical obstacle 2ft 3in (0.685m); trench 8ft 2in (2.489m); gradient 60 per cent.
History: Entered service with the Japanese Ground Self-Defence Force (JGSDF) in 1962. Being phased out of service, with some 400 remaining.

In appearance the Type 61 has a number of features of the American M47 medium tank, which the Japanese tested in small numbers in the early 1950s. The hull of the Type 61 is of all-welded construction, but the glacis plate can be removed for maintenance purposes. The driver is seated at the front of the hull on the

right. The turret is cast, with the commander and gunner on the right and the loader on the left. A stowage box is mounted at the rear of the turret bustle. The engine and transmission are at the rear of the hull. The Japanese have always favoured diesel engines as these have a number of advantages over petrol engines, including low fuel consumption and much-reduced fire hazard. The engine is air-cooled and turbocharged. The suspension is of the torsion-bar type and consists of six road wheels, with the drive sprocket at the front and the idler at the rear. There are three track-return rollers. The Type 61 is armed with a 90mm gun built in Japan, and there is a .3in machine-gun mounted co-axially with the main armament. The gun is elevated and traversed hydraulically, with manual controls for use in an emergency. An M2 Browning machine-gun is mounted on the commander's cupola for anti-aircraft defence and this can be aimed and fired from within the cupola. The tank can ford to a depth of 3ft 3in (0.99m) without preparation, but there is no provision for the installation of a schnorkel for deep fording operations. Some tanks have been provided with both infra-red driving lights and an infra-red searchlight for night operations. Compared with other tanks of the early 1960s, such as the Leopard and AMX-30, the Type 61 is undergunned, but it should be remembered that it was designed to meet Japanese rather than European requirements. The weight and size of the tank had to be kept within certain dimensions as the tank has to be able to be carried on Japanese railways, which pass through numerous narrow tunnels. There are three basic variants of the Type 61 MBT. The bridgelayer is called the Type 67 Armoured vehicle-Launched Bridge, and has a scissors-type bridge which unfolds over the forward part of the hull. This model weighs 81,499lb (37,000kg) and has a crew of three. Armament consists of a single .3in machine-gun. The recovery version is known as the Type 70 Armoured Recovery Vehicle. On this vehicle the turret is replaced by a small flat-sided superstructure. An 'A' frame is pivoted on this to lift tank components. A dozer blade is provided at the front of the hull. The ARV has a crew of four and a loaded weight of 77,094lb (35,000kg). Armament consists of a .3in and a .5in machine-gun and an 81mm mortar. Finally there is an Armoured Engineer Vehicle known as the Type 67. This weighs 77,094lb (35,000kg) and has a crew of four.

Left: The Type 61 was the first tank to be developed in Japan after the end of World War II and entered service with the Japanese Ground Self-Defence Force in 1962. It remains in service but has been supplemented by the Type 74 MBT, also developed by the Mitsubishi company.

Type 74 Main Battle Tank

Country of origin: Japan.
Crew: 4.
Armament: One L7 series 105mm gun; one 7.62mm machine-gun co-axial with main armament; one .5in anti-aircraft machine-gun, six smoke dischargers.
Armour: Classified.
Dimensions: Length (gun forward) 30ft 10in (9.41m); length (hull) 22ft 6in (6.85m); width 10ft 5in (3.18m); height (with anti-aircraft machine-gun) 8ft 10in (2.67m) at a ground clearance of 2ft 2in (0.65m).
Weight: Combat 83,776lb (38,000kg).
Ground pressure: 12lb/in² (0.85kg/cm²).
Engine: Mitsubishi 10ZF Model 21 WT 10-cylinder air-cooled diesel developing 750bhp at 2,200rpm.
Performance: Maximum road speed 33mph (53km/h); range 186 miles (300km); vertical obstacle 3ft 3in (1m); trench 8ft 10in (2.7m); gradient 60 per cent.
History: In service with Japanese Ground Self-Defence Force (JGSDF) in 1973. Production continues, with some 800 in service. ▶

Right: The first example of the Type 74 was called the STB-1. Features included an automatic loader for the 105mm main gun and a 12.7mm anti-aircraft machine-gun which could be aimed and fired by the commander from within the tank's turret.

Above: The Type 74 Main Battle Tank is manufactured by Mitsubishi Heavy Industries near Tokyo and is armed with the British-designed 105mm L7 type rifled tank gun which is manufactured in Japan under licence.

The Japanese realised in the early 1960s that the Type 61 would not meet its requirements for the 1980s, so in 1962 design work commenced on a new main battle tank. The first two prototypes, known as STB-1s, were completed at the Maruko works of Mitsubishi Heavy Industries in late 1969. Further prototypes, the STB-3 and the STB-6, were built before the type was considered ready for production. The vehicle entered production at the new tank plant run by Mitsubishi Heavy Industries at Sagamihara in 1973, and the first order was for 280 tanks. The Type 74 has not been exported as at the present time it is the policy of the Japanese government not to export arms of any type. The layout of the tank is conventional, with the driver at the front of the hull on the left and the other three crew members in the turret. The commander and gunner are on the right and the loader is on the left. The engine and transmission are at the rear of the hull. The suspension is of the hydro-pneumatic type and consists of five road wheels, with the drive sprocket at the rear and the idler at the front. There are no track-return rollers. The suspension can be adjusted by the driver to suit the type of ground being crossed. When crossing a rocky, broken area, for example, the suspension would be adjusted to give maximum ground clearance. This clearance can be adjusted from a minimum of 8in (20cm) to a maximum of 2ft 1½in (65cm). It can also be used to give the tank a tactical advantage: when the tank is on a reverse slope, the suspension can be lowered at the front and increased at the rear so that the main armament is depressed further than normal. The only other tank in service with this type of suspension is the Swedish S-tank, which has to have this type of suspension as the gun is fixed to the hull. This type of suspension was also used on the American T95 and the German/American MBT-70 tanks, but both these projects were cancelled. The Type 74 is armed with the British L7 series 105mm rifled tank gun, built under licence in Japan. A 7.62mm machine-gun is mounted co-axial with the main armament. The main gun has an elevation of +9° and a depression of −6° and using the hydropneumatic suspension an elevation of +15° and a depression

Above: One of the more interesting features of the Type 74 MBT is its hydro-pneumatic suspension which enables the driver to adjust the ground clearance to suit the type of terrain being crossed. In addition, when the tank is firing on a reverse slope the suspension can be raised at the front and lowered at the rear to give the 105mm gun is a depression angle of − 12.5° (normal depression angle is − 6°). Principle armament is an L7 series 105mm gun, backed up by a co-axial 7.62mm machine-gun. A 12.7mm machine-gun is mounted atop the turret for anti-aircraft purposes.

of − 12.5° can be obtained. The fire control system includes a laser rangefinder and a ballistic computer, both of which are produced in Japan. Some 51 rounds of 105mm ammunition are carried. Prototypes had an automatic loader, but this would have cost too much to install in production tanks. A .5in M2 anti-aircraft machine-gun is mounted on the roof. On the prototypes this could be aimed and fired from within the turret, but this was also found to be too expensive for production vehicles. Three smoke dischargers are mounted on each side of the turret. The tank is provided with infra-red driving lights and there is also an infra-red searchlight to the left of the main armament. The Type 74 can ford to a maximum depth of 3ft 3in (1m) without preparation, although a schnorkel enabling it to ford to a depth of 6ft 6in (2m) can be fitted. All tanks are provided with an NBC system. In designing the Type 74 MBT the Japanese sought, and managed, to combine the best features of contemporary tank design within a weight limit of 83,702lb (38,000kg). There is only one variant of the Type 74 at the present time, which is the Type 78 Armoured Recovery Vehicle; this is provided with a hydraulically-operated crane, winch and a dozer blade at the front of the hull. It has been produced in small numbers. The Type 74 chassis has also been used for the prototype AW-X twin 35mm SP anti-aircraft gun, production of which has yet to be authorised.

Type 75 155mm Self-propelled Howitzer

Country of origin: Japan.
Crew: 6.
Armament: One 155mm howitzer; one 12.7mm anti-aircraft machine-gun.
Armour: Classified.
Dimensions: Length (with armament) 25ft 6in (7.79m); length (hull) 21ft 9in (6.64m); width 10ft 1in (3.09m); height (turret roof) 8ft 4in (2.55m).
Weight: Combat 55,686lbs (25,300kg).
Ground pressure: 9.10lb/in² (0.64kg/cm²).
Engine: Mitsubishi 6ZF 6-cylinder diesel developing 450hp at 2,200rpm.
Performance: Road speed 29mph (47km/h); range 186 miles (300km); vertical obstacle 2ft 3in (0.7m); trench 8ft 2in (2.5m); gradient 60 per cent.
History: Produced 1977-88. 200 in service with Japanese Ground Self-Defence Force.

The first self-propelled artillery to be used by the Japanese Ground Self Defence Force in the post-Second World War period was introduced into service in 1965, when 30 105mm M52A1 and ten 155mm M4A1 self-propelled howitzers were procured from the United States. In 1967 development of a 105mm self-propelled howitzer commenced in Japan with Komatsu being responsible for the hull and the Japan Steel Works being responsible for the turret and main armament. This was eventually standardised as the Type 74 105mm self-propelled howitzer, but only 20 were built between 1975 and 1978 as it was decided to concentrate funding on the more effective 155mm weapon.

In 1969 the development of a 155mm self-propelled howitzer had started in Japan with Mitsubishi Heavy Industries being responsible for the turret and Nihon Seiko Jyo/Japan Iron Works being responsible for the turret and main armament. The first two prototypes were completed in 1971/72 and the vehicle was subsequently standardised as the Type 75 self-propelled howitzer. Production commenced shortly afterwards. Mitsubishi Heavy Industries manufacture the hull and also carry out final assembly and testing before delivering the complete system to the Army. In appearance the Type 75 is very similar to the American 155m M109A1 self-propelled howitzer, but the Japanese model has a slightly longer range.

The hull and turret of the Type 75 are of all welded alluminium construction with the driver seated at the front of the hull with the engine to his left and the turret at the very rear of the hull. Doors are provided in the rear of the hull for

Below: In appearance the Type 75 155mm SPH is similar to the American 155mm M109A1, but the Japanese weapon has a higher rate of fire (6rpm for three minutes) and a slightly longer range.

Above: The Type 75 155mm SP howitzer was designed by a group of companies, comprising Mitsubishi Heavy Industries, Nihon Seiko Jyo and the Japan Iron Works. Elevation of the 155mm howitzer ranges from −5° to +65°.

ammunition resupply purposes and there are also hatches and doors in the turret. The suspension is of the torsion bar type and consists of six rubber-tyred road wheels with the drive sprocket at the front and the last road wheel acting as the idler; there are no track return rollers.

Main armament consists of a long-barrelled 155mm howitzer which is provided with a double baffle muzzle brake and a fume extractor, and when travelling the howitzer is normally held in position by a travelling lock. The howitzer fires a Japanese HE projectile to a maximum range of 20,687 yards (19,000m), or an American projectile to a maximum range of 16,410 yards (15,000m). Elevation is from −5° to +65° and the turret can be traversed through a full 360°. Both elevation and traverse are hydraulic with manual controls provided for emergency use.

An unusual feature of the Type 75 is the loading system. In the rear of the turret are two drums, each of which holds nine projectiles, and these, together with the extendable loading tray and the power-operated rammer, enable 18 rounds to be fired in three minutes before the two drum magazines have to be reloaded. The latter can be accomplished from inside or outside the vehicle. A total of 28 155mm projectiles are carried plus the necessary bagged charges and fuses.

A 12.7mm machine-gun is pintle-mounted on the roof for anti-aircraft defence, and is provided with a small shield and a total of 1,000 rounds of ammunition. The Type 75 is fitted with an NBC system and infra-red night driving equipment and can ford to a depth of 4ft 3in (1.3m) without preparation.

Type 88 Main Battle Tank

Country of origin: South Korea.
Crew: 4.
Armament: One M68A1 105mm rifled gun; one 7.62mm co-axial machine-gun, one 12.7mm anti-aircraft machine-gun (gunner) and one 7.62mm machine gun (loader); 12 smoke dischargers (six on each side of turret).
Armour: Composite.
Dimensions: Length (including main armament) 31ft 9in (9.67m); length (hull) 24ft 6in (7.48m); width 11ft 10in (3.60m); height 7ft 5in (2.25m).
Weight: Combat 112,335lb (51,000kg).
Ground pressure: 12.23lb/in² (0.86kg/cm²).
Engine: MTU 871 Ka-501 12-cylinder water-cooled diesel engine developing 1,200hp at 2,600 rpm.
Performance: Road speed 40mph (65km/h); range 310 miles (500km); vertical obstacle 3ft 4in (1.0m); trench 9ft 0in (2.74m), gradient 60 per cent.
History: Entered production for the Republic of Korea Army in 1985. Some 300 in service in late 1991 out of an initial order for some 833.

Right: A view of an almost completed Type 88 MBT at Hyundai's Changwon manufacturing plant. 833 examples have been ordered.

Below: A pair of RoK Army Type 88s (also known as the K-1) partake in a live-firing exercise with their M68 105mm guns.

▶ The Republic of Korea (RoK) Army has traditionally used American equipment and its tank corps was equipped with a succession of such tanks, including the M47 and M48A5. It came, therefore, as something of a surprise when the RoK Army issued invitations in the mid-1970s for proposals for a new MBT, which was to be designed to South Korean specifications, with a view to production in South Korea. Several manufacturers submitted proposals and the Chrysler Defense (later General Dynamics Land Systems Division) design was selected in 1980. Two prototypes, designated XK-1, were built in considerable secrecy and sent to the Aberdeen Proving Grounds in Maryland, USA, for testing in 1983. Production started in 1984 at the Hyundai factory at Changwon under the designation Type 88 MBT and several tank battalions had been equipped with the new MBT by the time it was revealed to the public in late 1987. The total RoK Army requirement is believed to be about 800.

The Type 88 is of conventional design with the vulnerable parts of the hull constructed of British-designed and American-manufactured Chobham armour. There is a four-man crew, with the driver sitting in the front compartment on the left. The commander is seated in the turret on the right of the gun with the gunner in front of and below him, and the loader on the left of the gun.

The main armament is an M68A1 105mm rifled gun, another British design

Below: Moving forward to engage "enemy" forces in a training exercise, these RoK Army Type 88s bear high-visibility fabric panels to aid identification during the heat of battle.

built in the USA. The great majority of Western MBTs currently in production are armed with 120mm guns and it is a little surprising that the South Koreans should have opted for the older and smaller calibre weapon. However the 105mm is well-proven, very accurate and quite capable of defeating any tank operated by the North Koreans. In addition, the Rok Army's M48A5s are armed with the same weapon and there is an ammunition production facility in the country, producing all natures of 105mm ammunition including APFSDS. There are no known plans to up-gun the Type 88's primary armament to 120mm calibre.

There is the usual coaxial 7.62mm machine-gun. There are also two machine-guns mounted on the turret roof: the gunner has a 12.7mm weapon, while the loader's is of 7.62mm calibre. There are also 12 smoke grenade dischargers.

The rear-mounted engine is a German MTU 871 Ka-501, developing 1,200hp and giving the tank a power:weight ratio of 23.5hp/ton. The hybrid suspension is similar to that used in modern Japanese tanks, with the centre pair of road wheels having torsion bars and the front and rear pairs having hydropneumatic suspension. As with the Japanese tanks this enables the Type 88 to elevate its main gun to +20° and to depress it to -10°.

Two variants have been announced. One is an AVLB, for which the British firm of Vickers is building the scissors bridge and Hyundai the chassis, the latter being adapted from the Type 88. The other is an armoured recovery vehicle (ARV), for which Hyundai is preparing the design in cooperation with the German firm of Krupp-MaK, who build the Bergepanzer-2 and -3, which are the ARV versions of the Leopard 1 and 2, respectively.

SO-122 (2S1) (M-1974) Gvodzika 122mm Self-propelled Howitzer

Country of origin: Soviet Union.
Crew: 4.
Armament: One 2A31 122mm howitzer.
Armour: Hull 0.59in (15mm); turret 0.8in (20mm).
Dimensions: Length 23ft 10in (7.26m); width 9ft 4in (2.85m); height 8ft 11in (2.72m).
Weight: 34,611lb (15,700kg).
Engine: YaMZ-238V V-8 water-cooled diesel developing 240hp.
Performance: Road speed 38mph (62km/h); range 310 miles (500km); vertical obstacle 3ft 7in (1.1m); trench 9ft 9in (3m); gradient 60 per cent.
History: Entered service with the Soviet Army in 1971. In service in Algeria, Angola, Bulgaria, Czechoslovakia, Ethiopia, Hungary, Iraq, Libya, Poland, Syria Soviet Union and Yugoslavia.

The 122mm 2S1 self-propelled (SP) howitzer was one of two new Soviet SP artillery pieces to be introduced into service in the early 1970s, the other being the 152mm 2S3. The hull is of all-welded steel construction.

The layout of the M1974 is similar to that of the M1973 with the driver seated towards the front on the left, engine and transmission to the right and the turret well towards the rear of the hull. To the immediate front of the driver is a windscreen that can be covered by an armoured shutter hinged at the top when the vehicle is in a combat area, and over the top of the driver's position is a single-piece hatch cover that opens to the rear. The commander is seated on the left side of the turret and has a cupola that can be traversed through a full 360° and is provided with a single-piece hatch cover that opens to the front, forward of which are periscopes and an infra-red searchlight that can be operated by the commander from within the turret. The gunner is seated forward and below the commander and the loader on the right side of the turret, with a single-piece hatch cover that opens forwards.

Main armament consists of a a modified version of the D-30 122mm towed

Below: The SO-122 (2S1) armed with a 122mm howitzer.

Above: The Polish Army is one operator of the SO-122.

howitzer, designated 2A31. It is fitted with a large double baffle muzzle brake and a fume extractor. The OF-462 high-explosive projectile weighs 47.88lb (21.5kg), with a maximum muzzle velocity of 2,264ft/s (690m/s) to a maximum range of 16,738 yards (15,300m). A rocket-assisted projectile has a range of 23,950 yards (21,900m). The gun can also fire a HEAT-FS round weighing 47.7lbs (21.63kg), which will penetrate 18-in (460mm) armour at a range of 1,094 yards (1,000m). A total of 40 rounds of ammunition are carried and maximum burst rate of fire is eight rounds per minute, with a capability of 70 rounds in the first hour of firing. The ordnance has full 360° traverse and elevation limits are +70° to −3°.

The 2S1 is fully amphibious, being propelled in the water by its tracks at a speed of 3mph (4.5km/h). Unusually, among Soviet amphibious armoured fighting vehicles, it is not fitted with a trim vane on the glacis plate. However, it is fitted with a full NBC system.

Like much Soviet equipment, this SP howitzer has a variety of designations, which may confuse the uninitiated. To the Soviet Army it is the SO-122, with the official nickname of *Gvodzika* (Carnation). To Soviet industry its production code is 2S1, while NATO refers to it as the M-1974, that being the year of its first public appearance.

SO-152 (2S3) Akatsiya (M-1973) 152mm Self-propelled Gun/ Howitzer

Country of origin: Soviet Union.
Crew: 4.
Armament: One 152mm howitzer; one 7.62mm anti-aircraft machine-gun.
Armour: Hull 0.59in (15mm); turret 0.8in (20mm).
Dimensions: Length (gun forwards) 27ft 6in (8.40m); width 10ft 6in (3.20m); height 9ft 2in (2.80m).
Weight: 66,137lb (30,000kg).
Engines: V-59 V-12 water-cooled diesel developing 520hp.
Performance: Road speed 38mph (62km/h); range 310 miles (500km); vertical obstacle 3ft 7in (1.1m); trench 8ft 2in (2.5m); gradient 60 per cent.
History: Entered service with the Soviet Army in 1971. In service in Hungary, Iraq, Libya, Soviet Union and Syria.

The SO-152 was introduced in the early 1970s on the scale of 18 SPs per division, although this was later increased to 24 per division. The design of the SO-152 is similar to that of most other SPs, with the driver and transmission at the front of the hull and the large turret towards the rear. The driver is at the front on the left and has a single-piece hatch hinged at the rear.

The large turret has a single hatch in the right side and a cupola for the vehicle commander is provided on the left side of the turret roof. The commander's cupola has a single-piece hatch cover that opens to the rear and is provided with vision devices; a 7.62mm machine-gun is mounted on the forward part of the cupola and it is possible that this can be aimed and fired from within the vehicle. The indirect sight is mounted in the turret roof forward of the commander's position with the direct sight being mounted to the left of the main armament. An ammunition resupply hatch is provided in the rear of the hull but, unlike in the American M109, there are no spades at the rear of the hull to absorb recoil when the weapon is fired. The suspension is of the torsion bar type and consists of six dual rubber-tyred road wheels with the drive sprocket at the front, idler at the rear and four track return rollers.

Main armament is the 2A33 152mm gun/howitzer, which has been developed from the towed 152mm D-20. The ordnance is fitted with a double baffle muzzle brake and a fume extractor. When travelling, the ordnance is held in place by

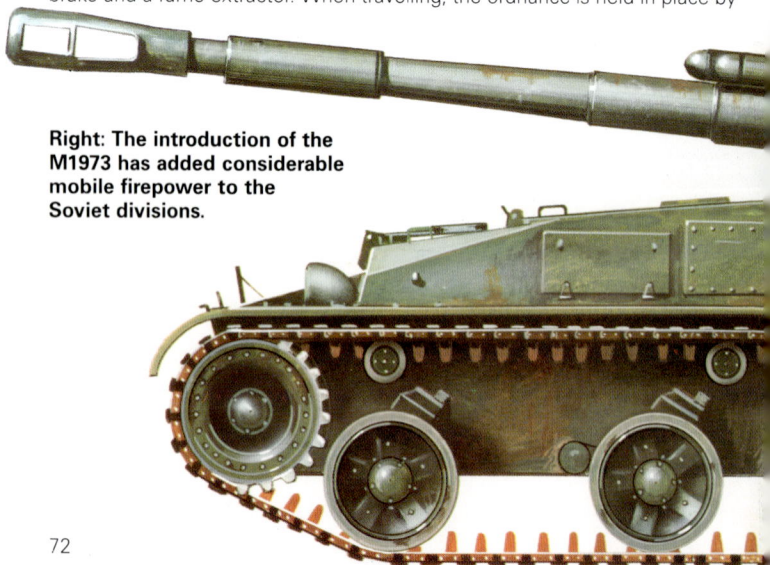

Right: The introduction of the M1973 has added considerable mobile firepower to the Soviet divisions.

Above: The 152mm SO-152 self-propelled gun/howitzer was seen in public for the first time during the November 1977 Moscow parade.

a travel lock which folds back onto the glacis plate when not in use. The ordnance has elevation limits of +60° to −3° and can be traversed through a full 360°. The main projectile is a high-explosive fragmentation round, designated OF-540, which weighs 95.9lbs (43.5kg) and has a maximum range of 20,230 yards (18,500m). There is also a rocket-assisted projectile with a range of 26,250 yards (24,000m). A fin-stabilised, high explosive, anti-tank round is also available. A total of 40 rounds is carried.

The maximum firing rate is four rounds per minute, although the sustained rate is two rounds per minute. The SO-152 can ford to a depth of 4ft 11in (1.5m) without preparation, but unlike the majority of Soviet armoured fighting vehicles (AFV) is not amphibious. There is a full NBC system and night-driving equipment is provided.

As with other Soviet AFVs this equipment has a variety of designations. The Soviet Army describes it as the SO-152 *Akitsya* (Acacia), while to Soviet industry it is the 2S3. NATO uses the designation M-1973, based on the year it was first seen in public.

SM-240 (2S4) Tyulpan (M-1975) 240mm Self-propelled Mortar

Country of origin: Soviet Union.
Crew: 4.
Armament: One 240mm mortar.
Dimensions: (estimated) Length (including baseplate) 28ft 4in (8.63m); length (chassis) 23ft 11in (7.05m); width 10ft 6in (3.23m); height 9ft 2in (2.80m).
Weight: (estimated) 72,750lb (33,000kg).
Engine: Water cooled diesel developing 450hp.
Performance: (estimated) Road speed 31mph (50km/h); road range 310 miles (500km).
History: Entered service with the Soviet Army in 1975. In service in Czechoslovakia, Iraq, Lebanon and Soviet Union.

This unusual weapon is by far the largest self-propelled (SP) mortar in service today. It is not, however, the largest SP mortar ever, that record belonging to a giant German device used in the siege of Sevastopol in World War II.

Below: An artist's impression of the SM-240 in travelling mode, with the mortar barrel facing forwards. When set to fire, the barrel is rotated through some 270° and faces towards the rear.

The Soviet designers have used their usual skill to match two existing weapons and produce an excellent new weapon system. The chassis is based on that used for the GMZ tracked minelayer and is similar to that used for the new 2S3 152mm SP gun, but with a somewhat stronger suspension to carry the very heavy mortar.

The mortar itself is a fairly standard version of the M-240 240mm towed, breech-loading mortar. The mortar is mounted on a baseplate which is hinged on the rear of the vehicle. To come into action the vehicle faces away from the line of fire and the mortar is hydraulically lifted through some 270° until the baseplate, which remains attached to the vehicle, is firmly embedded in the ground. Once in position the mortar can be elevated between + 45° and + 80°, and can be traversed up to 8° left or right.

The mortar can fire high-explosive (HE) and concrete-piercing rounds. It was also capable of firing nuclear and chemical rounds, although these have both now been withdrawn. The F-864 HE round weighs some 287lb (130kg) and can be fired to a maximum range of 10,610 yards (9,700m). It is also reported that there may be an extended-range round (possibly rocket-assisted) with a range of 21,870 yards (20,000m). The rounds are loaded into the breech with the assistance of a powered device, but it appears that the crew must work outside the vehicle during an engagement.

Some 500 of these weapons are in service with the Soviet Army and a smaller number with the Czechoslovak Army. It is also in service with Iraq, which country is believed to have passed some on to the Christian Militia in Lebanon. The Soviet Army designation is SM-240, with the official nickname of *Tyulpan* (Tulip Tree).

SO-203 (2S7) (M-1975) 203mm Self-propelled Gun

Country of origin: Soviet Union.
Crew: 4.
Armament: One 203mm gun.
Dimensions: Length 42ft 0in (12.81m); width 10ft 6in (3.23m); height 9ft 2in (2.80m).
Weight: 88,183lb (40,000kg).
Engine: Water-cooled diesel developing 450hp.
Performance: Road speed 31mph (50km/h); road range 310 miles (500km).
History: Entered service with the Soviet Army in 1975. In service in Czechoslovakia and Soviet Union.

The Soviet Army retained towed artillery in its front-line units long after Western armies had converted to self-propelled (SP) equipment. Once the Soviets began to convert, however, in the early 1970s the process was very quick and some excellent equipment has since been fielded.

Below: The SO-230 entered service with the Soviet Army in the late 1970s and is still the largest full-tracked self-propelled gun in the world. Note the travelling lock for the 203mm gun.

The SO-152 was the first to appear, followed by the SO-122 a year later. Next came this 203mm weapon, which is now in service with the Soviet and Czechoslovakian armies. The chassis is very large with a length of 34.45ft (10.51m), although its width of 10ft 6in (3.23m) is limited by the requirement to transport it on a standard railroad low-loader. Most Soviet SPs are mounted on chassis developed from those used for armoured personnel carriers or missile systems, but the SO-203 chassis appears to be unique. It has seven road wheels, with an amoured cab mounted at the very front of the vehicle, and can carry a full crew of four men. The engine is mounted behind the cab and drives the track through forward-mounted sprockets. There is a large hydraulically-operated spade at the rear of the vehicle.

The gun is mounted at the rear, but there is no gun-housing to provide cover for the crew. The 203mm ordnance has a range estimated to be 41,000 yards (37,500m) with standard HE rounds and 54,680 yards (50,000m). Maximum firing rate is two rounds per minute, but with a sustained rate of one round every two minutes. The gun fires high-explosive and concrete-piercing rounds. It was also capable of firing nuclear and chemical rounds, but these have been withdrawn following recent arms limitation agreements.

The projectile and cartridge cases for this weapon are large and heavy. A few rounds will be carried on the SP, but there is certain to be an ammunition supply vehicle associated with each weapon. Most photographs show a wheeled trolley, usually stowed on top of the vehicle behind the cab, which is almost certainly used to convey rounds from the resupply vehicle to the rear of the SO-203, where they are picked up by the semi-automatic loader.

ASU-85 Self-propelled Anti-tank Gun

Country of origin: Soviet Union.
Crew: 4.
Armament: One 85mm gun; one PKT 7.62mm machine-gun co-axial with main armament; one 7.62mm anti-aircraft machine-gun.
Armour: 0.39in-1.56in (10mm-40mm).
Dimensions: Length (with armament) 27ft 10in (8.49m); length (hull) 19ft 8in (6m); width 9ft 2in (2.8m); height 6ft 11in (2.1m).
Weight: 34,171lb (15,500kg).
Ground pressure: 6.25lb/in² (0.44kg/cm²).
Engine: Model V-6 six-cylinder inline water-cooled diesel developing 240bhp at 1,800rpm.
Performance: Road speed 27.3mph (44km/h); range 162 miles (260km); vertical obstacle 3ft 8in (1.1m); trench 9ft 2in (2.8m); gradient 70 per cent.

Above: The 85mm gun of the ASU-85 has a fume extractor and a double baffle muzzle brake. The APHE projectile weighs 20.5lb (9.3kg), has a muzzle velocity of 2,598ft/s (792m/s) and will penetrate 102mm of armour at a range of 1,093 yards (1000m); the HVAP projectile will penetrate 130mm of armour at this range.

History: Entered service with the Soviet Army in 1961. Production completed in 1964. No variants and serves with Soviet airborne forces only.

The ASU-85 (ASU is the abbreviation for *Aviadezantnaya Samochodnaya Ustanovka*, and the 85 refers to the size of the main armament) is normally transported in the Antonov An-12 "Cub" transport aircraft and can be air-dropped. For air-dropping the vehicle is mounted on a platform to which are attached a number of parachutes. Just before the platform reaches the ground a number of retro-rockets are fired to reduce the platform's velocity so that no damage occurs. The ASU-85 has a hull of all-welded steel construction which varies in thickness from 0.4in (10mm) on the hull roof to 1.56in (40mm) on the glacis and mantlet. The fighting compartment is at the front, with the engine and transmission at the rear. Many components of the ASU-85 are taken from the PT-76 amphibious light tank family. The crew consists of commander, gunner, loader and driver, the last being seated at the front of the vehicle on the right side. The ASU-85 has torsion-bar suspension and a total of six road wheels with the idler at the front and the drive sprocket at the rear, but does not have any track return rollers. The 85mm gun is provided with a double baffle muzzle-brake ▶

Left: Main armament of the ASU-85 consists of an 85mm gun mounted in the forward part of the hull. This fires fixed APHE, HE and HVAP rounds and has an elevation of +15°, depression of −4° and total traverse of 12°.

Below: Mounted over the main armament of the ASU-85 at the rear is an infra-red searchlight; a smaller one is mounted forward of the commander's position to the right of the 85mm gun and this can be operated safely from within the vehicle.

▶ and a fume extractor, and is mounted slightly offset to the vehicle's left; traverse is a total of 12° and elevation from − 4° to + 15°. A PKT 7.62mm machine-gun is mounted co-axially with the main armament. A total of 40 rounds of 85mm ammunition is carried, including HE, APHE and HVAP. The HE projectile weighs 20.9lb (9.5kg) and has a muzzle velocity of 2,598ft/s (792m/s), the APHE projectile weighs 20.5lb (9.3kg) and also has a muzzle velocity of 2,598ft/s, and the HVAP projectile weighs 11 02lb (5kg) and has a muzzle velocity of 3,379ft/s (1,030m/s). The APHE round will penetrate 4in (102mm) of armour at a range of 1,093 yards (1,000m), whilst the HVAP round will penetrate 5.12in (130mm) of armour at a similar range. The ASU-85 is fitted with an NBC system. Infra-red driving lights are fitted and there is an infra-red searchlight over the main armament and another in front of the commander's hatch, the last controllable from within the vehicle. The vehicle does not have any amphibious capability, although it can ford to a depth of 3ft 8in (1.1m) without preparation. Two fuel drums can be attached to the rear of the hull to increase operational range.

The ASU-85 is now well over 25 years old and the performance of the 85mm gun against modern MBTs must be marginal, but the type remains in full service with Soviet airborne forces. There is a battalion of 31 vehicles in each airborne division: one is at battalion headquarters and the remainder are in three companies, each of 10 vehicles.

Left: An ASU-85 being unloaded from a Soviet Air Force Antonov An-12 "Cub" transport aircraft. This self-propelled anti-tank gun is issued to Soviet airborne forces on the scale of 31 per division.

Below: The ASU-85 was developed in the late 1950s and was seen in public for the first time during a parade held in Moscow in 1962. It uses many parts of the PT-76 light tank family but, unlike this vehicle, is not amphibious.

PT-76 Light Amphibious Tank

Country of origin: Soviet Union.
Crew: 3.
Armament: One 76.2mm gun; one SGMT 7.62mm machine-gun co-axial with main armament.
Armour: 0.55in (14mm) maximum.
Dimensions: Length (gun forward) 25ft (7.62m); length (hull) 22ft 8in (6.91m); width 10ft 4in (3.14m); height 7ft 2in (2.19m).
Weight: Combat 30,865lb (14,000kg).
Ground pressure: 6.8lb/in² (0.48kg/cm²).
Power to weight ratio: 17.1hp/t.
Performance: Maximum road speed 27.34mph (44km/h); water speed 6.2mph (10km/h); range 162 miles (260km); vertical obstacle 3ft 8in (1.1m); trench 9ft 2in (2.8m); gradient 60 per cent.
History: Entered service with the Soviet Army in 1952; production of some 7,000 completed in late 1960s. In service with: Afghanistan (60), Algeria (50), Angola (50), Benin (20), Cambodia (10), Congo (3), Cuba (60), Egypt (15), Finland (15), Guinea (20), Guinea-Bissau (20), India (100), Indonesia (41), Iraq (100), Laos (25), Madagascar (12), Nicaragua (22), Poland (100), Soviet Union (1,200), Vietnam (150), Yugoslavia (13) and Zambia (30).

The PT-76 (*Plavaushiy Tank*) is based on the *Pinguin* cross-country vehicle. Since it entered service with the Soviet Army in 1952, it has been exported to many countries and has seen combat in Africa, the Middle East and the Far East. It has a hull of all-welded steel construction. The driver is seated at the front of the hull, with the commander/gunner and loader in the turret, and the engine and transmission at the rear of the hull. The PT-76 is armed with a 76.2mm gun, this having an elevation of + 30° and a depression of − 4° A 7.62mm SGMT machine-gun is mounted co-axially with the main armament. 40 rounds of 76.2mm and 1,000 rounds of 7.62mm ammunition are carried. The most outstanding feature of the PT-76 is its amphibious capability. It is propelled in the water by two water-jets, one in each side of the hull, with their exits in the hull rear. Before entering the water a trim vane is erected at the front of the hull and the driver's centre periscope is raised so that he can see over the top of the trim vane. The PT-76 has been built in large numbers and its basic chassis has been used for a whole family of other armoured vehicles. A modified version

Above right: PT-76 Model 2s show their amphibious capabilities: note trim vane erected, driver's periscope extended and schnorkel on the rear of the turret.

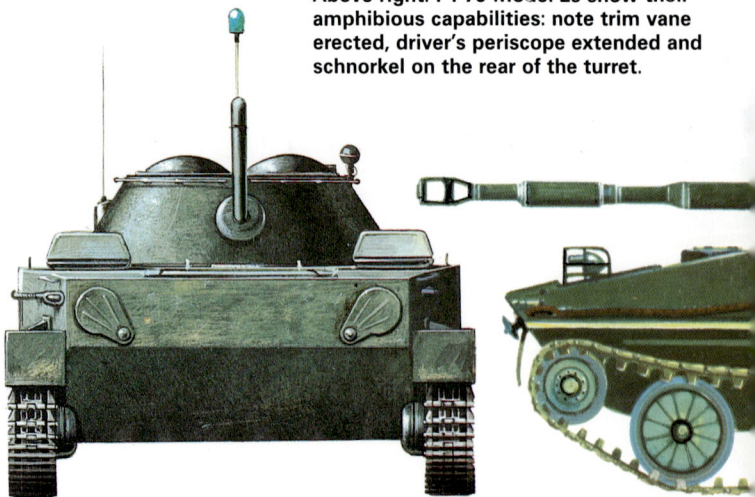

has been built in China as the Type 63. This has a similar hull to the PT-76, but has a new turret mounting an 85mm gun and a co-axial 7.62mm machine-gun; there is also a 12.7mm anti-aircraft machine-gun on the roof. Although now some 40 years old, the PT-76 is still a useful vehicle in the reconnaissance role.

Left: Side and rear drawings of a PT-76 Model 2 showing covered waterjet outlets at rear of hull and inlets on hull sides towards rear.

SO-120 (2S9) Self-propelled 120mm Howitzer/Mortar

Country of origin: Soviet Union.
Armament: One 120mm howitzer/mortar.
Dimensions: Length 19ft 9in (6.02m); width 8ft 7in (2.63m); height 7ft 6in (2.29m).
Weight: Combat 17,621lb (8,000kg).
Engine: One 5D20 diesel developing 300hp.
Performance: Road speed 37mph (60km/h); range 310 miles (500km); gradient 60 per cent.
History: Entered service with the Soviet Army in 1984. Production complete. In service with Soviet Parachute troops, Soviet Army, Soviet Marines and Afghan Army.

This unusual combat vehicle was developed specifically for the elite airborne forces. The 120mm-calibre, 5ft 11in (1.8m)-long mortar tube is mounted in a conventional turret, which can traverse 35° either side of straight ahead. What makes the vehicle so unusual, however, is that the tube can be elevated up to + 80°, which means that it can be used as a mortar, firing bombs out to a range of some 9,600 yards (8,800m). There are three types of mortar bomb: high-explosive, white phosopherous and smoke. The loading system is semi-automatic, in which the loader places a round manually on a feed tray and then presses

a button, following which the round is automatically rammed and the breech closed. With a well-trained crew, a maximum firing rate of about eight rounds per minute is possible.

The tube can also be used in the direct fire mode, firing a HEAT projectile with an effective range of some to 2,000 yards (1,830m). This round is reported to be capable of defeating armour of up to 2.36in (6cm) thickness. A total of 60 rounds of all types are carried. No machine-guns are mounted.

The normal height of the vehicle for travelling is 7ft 6in (2.29m). The suspension is lowered for firing to provide a more stable platform, reducing height to 6ft 3in (1.90m).

There is a crew of four. The driver sits at the centre of the glacis plate and there is a two-man turret, with the gunner on the left and the loader on the right. Another curious feature of the vehicle is that the commander is not in the turret, but sits under a hatch to the left of the driver in the front of the vehicle.

The vehicle is fully amphibious. There is a splash vane on the glacis plate and a bilge pump is fitted. Propulsion is by means of two water jets at the rear, giving a speed of some 6mph (9km/h) in still water.

The SO-120 *Anona* (Anemone) can be transported in any Soviet medium or heavy cargo aircraft. It can also be air-dropped using the standard Soviet heavy-drop platform with multiple parachutes and retro-rockets.

Altogether this is a remarkable fire support vehicle, showing once again the ability of Soviet military designers to get "a quart into a pint pot". In combat the SO-120 can arrive with the leading elements in a parachute assault and give both indirect and direct fire support.

Left: The modest dimensions of the SO-120 can be appreciated in this photo, showing two of the four-man crew. Its modest size allows the SO-120 to be airlifted by several Soviet transport aircraft — a very great advantage when the emphasis is on providing fire support for advancing troops both quickly and effectively.

T-55 Main Battle Tank

Country of origin: Soviet Union.
Crew: 4.
Armament: One 100mm gun; one SGMT 7.62mm co-axial machine-gun, two SGMT 7.62mm anti-aircraft machine-guns.
Armour: 5.9in (150mm) maximum.
Dimensions: Length (including main armament) 29ft 6in (9.0m); length (hull) 21ft 2in (6.45m); width 10ft 9in (3.27m); height 7ft 10in (2.4m).
Weight: Combat 91,410lb (41,500kg).
Engine: 12-cylinder four-stroke water-cooled diesel engine developing 630hp at 2,000 rpm.
Performance: Road speed 31mph (50km/h); road range with two additional 200l fuel tanks 340 miles (545km); vertical obstacle 2ft 8in (0.8m); trench 8ft 10in (2.7m); gradient 60 per cent.
History: T-54 entered service with Soviet Army in 1947, T-55 in 1960. Over 60,000 T-54/T-55s produced in Soviet Union, 2,000 in Poland and 3,000 in Czechoslovakia, of which some 47,280 are still in service with 47 armies. Production of T-55 ended in Soviet Union in 1981. Copy of T-54 produced in China (see Type 59).
(Data in specification box applies to T-55 AM2B.)

Over 65,000 T-54/T-55 MBTs have been produced in the Soviet Union,

Czechoslovakia and Poland, plus a further 7,000 Type 59s in China. This beats the 50,000-unit production figures of the Soviet T-34 by a handsome margin, making the T-54/T-55 the most widely-used tank of all time; a record which is unlikely to be broken in the future.

During World War II, the Soviet Army's main tank design was the T-34, which mounted an 85mm gun and is generally accepted to have been the best all-round tank of the war. The T-34 was developed into the T-44, which was produced in small numbers, but turned out to be an interim design, as the much improved T-54 appeared in 1947 and became the Warsaw Pact's first standard MBT. The

T-54 is armed with the D-10T 100mm gun, firing APHE, HEAT and HE rounds, and is capable of an elevation of + 17° and a depression of -4°, the latter being significantly less than in Western MBTs. The turret is virtually hemispherical in shape, which gives good ballistic protection, but makes it somewhat cramped inside by Western standards.

The T-55 appeared in 1960 and incorporated many improvements, including a more powerful engine. It uses the same 100mm main gun and the first production model also retained the bow-mounted machine-gun although this was deleted from the T-55A onwards.

There are many modified versions of the T-54/-55 series and sophisticated retrofit programmes have been undertaken by the Israeli Army, the British firm of Royal Ordnance and the US firm of Teledyne Continental. The Soviet Union also continues to update versions in service with the Soviet Army, of which one of the most recent is the T-55 AM2B. This has a new turret, appliqué armour on both turret and hull, a new and more powerful engine, much-improved electronics and vision devices, and the same tracks as used on the T-72.

This new version is designed to fire the 9K116 anti-tank guided missile from its 100mm main gun. This missile has a maximum range of some 4,400 yards (4,023m), using a semi-automatic guidance system which requires the gunner to keep his sight on the target until impact. The shaped-charge warhead on the missile is small in diameter, which limits its effectiveness against the latest western tanks which use Chobham-type armour, although it would be very effective against older tanks.

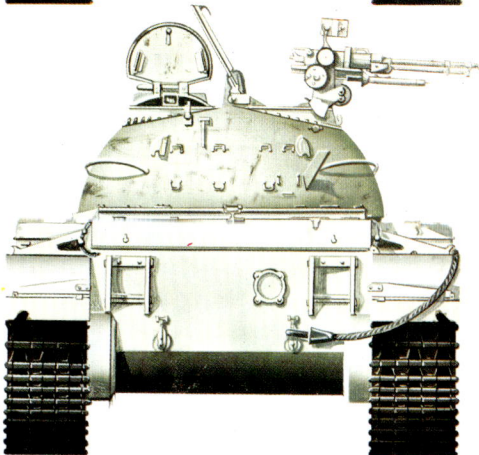

Above: An Egyptian Army T-55 after the fitment of a 105mm gun in the UK.

Right: Front and rear views of a T-54 tank while serving with the Egyptian Army.

T-62 Main Battle Tank

Country of origin: Soviet Union.
Crew: 4.
Armament: One U-5TS 115mm gun; one PKT 7.62mm PKT machine-gun co-axial with main armament; one DShK 12.7mm anti-aircraft machine-gun (optical).
Armour: 0.79in-9.52in (20mm-242mm).
Dimensions: Length (overall) 30ft 7in (9.33m); length (hull) 21ft 9in (6.63m); width 11ft (3.35m); height (without anti-aircraft machine-gun) 7ft 10in (2.4m).
Weight: Combat 88,200lb (40,000kg).
Ground pressure: 11.8lb/in² (0.83kg/cm²).
Engine: Model V-55-5 12-cylinder water-cooled diesel engine developing 580hp at 2,000rpm.
Performance: Road speed 31mph (50km/h); range (without additional fuel tanks) 280 miles (450km); vertical obstacle 2ft 8in (0.8m); trench 9ft 2in (2.8m); gradient 60 per cent.
History: Entered service with the Soviet Army in 1963. Produced in Soviet Union (c.20,000), Czechoslovakia (1,500) and North Korea (c.2,000). Production has ended. Approximately 19,000 remain in service in 17 armies.

The T-62 was developed in the late 1950s as the successor to the earlier T-54/T-55 series, and was first seen in public in May 1965. In appearance it is very similar to the earlier T-54. It does, however, have a longer and wider hull, a new turret and main armament, and can easily be distinguished from the T-54 as the latter has a distinct gap between its first and second road wheels, whereas the T-62's road wheels are more evenly spaced, and the T-62's gun is provided with a bore evacuator. The hull of the T-62 is of all-welded construction with the glacis plate being 4in (10cm) thick. The turret is of cast armour, and this varies in thickness from 6.7in (17cm) at the front to 2.4in (6cm) at the rear. The driver is seated at the front of the hull on the left side, with the other three crew members in the turret, the commander and gunner on the left and the loader on the right. The engine and transmission are at the rear of the hull. The suspension is of the well-tried torsion bar type, and consists of five road wheels with the idler at the front and the drive sprocket at the rear. The U-5TS gun is of the smoothbore type, and has an elevation of +17° and a depression of −4°. A PKT 7.62mm machine-gun is mounted co-axially with the main armament. When the T-62 first entered service it did not have an anti-aircraft machine-gun, but many T-62s have since been provided with the standard DShK 12.7mm weapon which is mounted on the loader's cupola, T-62s thus fitted being designated T-62A. Three types of ammunition are carried — High Explosive, Fin-Stabilised Armour-Piercing Discarding Sabot (FSAPDS) and High Explosive Anti-Tank (HEAT). The FSAPDS round has a muzzle velocity of 5,512ft/s (1,689m/s) and an effective range of 1,749 yards (1,600m). When this round is fired the sabot (the disoposable "slipper" around the projectile) drops off after the round has left the barrel and the fins of the projectile unfold to stabilise the round in flight. According to Israeli reports,

Right: Standard equipment on the T-62 includes an NBC system and night vision equipment including infra-red driving light, infra-red searchlight to the right of the main armament and infra-red search-light on the commander's cupola that can be operated from within the turret.

Above: A column of T-62s trundles past a Soviet Army motor-cycle reconnaissance team on exercise in the Soviet Union.

this round will penetrate 11.8in (30cm) of armour at a range of 1,094 yards (1,000m). The 115mm round is manually loaded, but once fired the gun automatically returns to a set angle at which the empty cartridge case is ejected from the breech, after which it moves onto a chute and is then thrown out through a small hatch in the turret rear. There are three variants of the T-62: the T-62M is an improved MBT, the T-62K is a command tank and the M1977 is an armoured recovery vehicle. Some 19,000 T-62s remain in service around the world and many are being upgraded by the addition of new armour, tracks, sideskirts and guns.

T-64 Main Battle Tank

Country of origin: Soviet Union.
Crew: 3.
Armament: One 2A26 125mm gun; one PKT 7.62mm co-axial machine-gun; one 12.7mm anti-aircraft machine-gun.
Armour: Classified.
Dimensions: Length (including main armament) 32ft 6in (9.90m); length (hull) 24ft 5in (7.45m); width 15ft 3in (4.64m); height 7ft 3in (2.20m).
Weight: Combat 92,512lb (42,000kg).
Engine: 5DTF, 5-cylinder opposed piston liquid-cooled diesel engine developing 750hp at 2,000rpm.
Performance: Road speed 47mph (75km/h); road range 250 miles (400km); vertical obstacle 2ft 8in (0.8m); trench 8ft 10in (2.7m); gradient 60 per cent.
History: T-64 entered service with Soviet Army in 1980 and serves only with the Soviet Army. Production has ended.
(Data in Specification box applies to T-64B.)

In the 1960s the Soviet Army built prototypes of a tank known as the M1970 MBT. This tank was similar to the T-62 and was armed with the same 115mm smoothbore gun, but differed in having an entirely new suspension system, with six small roadwheels. The M1970 was not put into production, but a development then appeared, designated the T-64, which entered service with the Soviet Army in the late 1960s. This has a similar hull and suspension to the M1970, but with a new turret mounting a new 125mm smoothbore gun, which is fed from an automatic loader.

The driver of the T-64 sits in the front of the vehicle in the centre. The other two crewmen are in the turret, with the commander on the right of the gun and the gunner on the left. The gun is the 2A26 125mm smoothbore, which has been the largest-calibre tank gun in service in any army since its appearance in the T-64. The 2A26 has a vertical ammunition stowage system for its automatic loader, which was reported to have given some trouble in its early years of service, with a number of crewmen being seriously injured. There is an infra-red searchlight

Below: The T-64B's 2A26 125mm smoothbore gun can fire the same ammunition as the T-72, and the AT-8 "Songster" missile.

Above: The T-64 MBT entered service in 1980 and has remained exclusive to the Soviet Army ever since. Production has ended.

mounted on the left side of the turret. There are 12 smoke-grenade launchers mounted individually on the forward face of the turret. There is also a thermal sleeve for the main gun, and hinges for track guards which are constructed of thin metal plate.

The suspension has six small, dual roadwheels mounted on hydropneumatic arms, a most unusual arrangement in the Soviet Army, which has used large roadwheels and torsion bars since the Christie tanks of the 1930s. There are two thin snorkels for fording operations, one for the gunner's periscope mount, the other for the engine air intake.

The T-64K is a minor variant of the basic T-64 designed for use as a command post tank, for which role it carries extra radios, slightly reducing the number of rounds of 125mm ammunition carried. A telescopic mast is carried on the outside of the hull, which is erected whenever the tank is at the halt.

Three major versions of the T-64 have been identified. The first was the initial production version described above. Next came the T-64A (designated T-64 M1981/1 by the US Army) which has a number of minor modifications. The T-64B then appeared in the early 1980s; some may have been new-builds but most appear to have been converted T-64As. The T-64B is fitted with the 2A46 125mm smoothbore gun, the same as that fitted to the T-80, although it appears to retain the same type of automatic loader as that in the T-64A. The rounds carried are APFSDS or HEAT-FS, but this tank can also launch an anti-tank guided-weapon, which is designated AT-8 "Songster" by NATO.

The AT-8 is propelled out of the barrel by a boost motor and the main motor then cuts in and powers it to the target at a speed of some 1,640ft/s (500m/s). Maximum effective range is about 4,400 yards (4,000m) and the HEAT warhead will penetrate about 3in (76mm) of steel armour, although its effect against modern ceramic and reactive armours is probably much less.

The T-64B is fitted with a laser rangefinder and additional sensors which are used to acquire helicopter targets prior to engagement with the AT-8 missile. The T-64B is also fitted with mountings for 111 blocks of bolt-on reactive armour, which cover the tank's glacis plate, hull sides and most of the turret. The blocks on the turret necessitate the smoke-grenade dischargers being moved to the rear of the turret abreast the commander's hatch.

As far as is known no T-64s have ever been exported and the type appears to have served only with the tank divisions in the Western military districts of the Soviet Union, with the Soviet forces in Hungary and in the Group of Soviet Forces Germany (now designated the Western Group of Forces).

T-72 Main Battle Tank

Country of origin: Soviet Union.
Crew: 3.
Armament: One 2A46 125mm gun; one PKT 7.62mm co-axial machine-gun; one NSVT 12.7mm anti-aircraft machine-gun.
Armour: Classified.
Dimensions: Length (including main armament) 30ft 4in (9.24m); length (hull) 22ft 10in (6.95m); width 15ft 7in (4.75m); height 7ft 10in (2.37m).
Weight: Combat 90.310lb (41,000kg).
Ground Pressure: 11.80lb/in² (0.83kg/cm²).
Engine: V-12 diesel engine developing 780hp at 2,000 rpm.
Performance: Road speed 50mph (80km/h); road range 300 miles (483km); vertical obstacle 2ft 8in (0.85m); trench 8ft 10in (2.7m); gradient 60 per cent.
History: Entered service with Soviet Army in 1973; in production in the Soviet Union, Czechoslovakia, Poland, India, Iraq, Romania and Yugoslavia. In service in Algeria (200), Bulgaria (200), Cuba (50), Czechoslovakia (850), Finland (80), Hungary (200), India (750), Iraq (500), Kuwait (?), Libya (180), Poland (400), Romania (100), Soviet Union Syria (1,000) and Yugoslavia (300 +).

Below: The T-72 MBT is in widespread service with the Soviet Army and a host of other armies around the world. It has a three-man crew and an automatic loader, with each round of ammunition having its charge and projectile loaded separately.

Although the T-72 bears an external resemblance in certain respects to the T-64 it was, in fact, developed at a quite different design bureau. It entered production in 1971 and was in wide-scale service by 1973, although it was not reported publicly by Western experts until 1977. As in all recent Soviet MBTs the driver is seated centrally under a well-sloped glacis plate, which has transverse ribs and a splashboard. The other two crew members are seated in the turret, the commander on the right and the gunner on the left. All T-72s built for use by armies of the (former) Warsaw Pact have an interior lining of a special synthetic, lead-based material. This is intended to provide some degree of protection against two of the products of a nuclear explosion: for the crew against neutron radiation and for the electronic equipment against electro-magnetic pulses.

There is an automatic loader, which, unlike that in the T-64, has a horizontal feed system. Except in the earliest version, the main gun is the entirely new 2A46 125mm smoothbore, which is fitted with a light alloy thermal sleeve and a fume extractor, and fires three types of ammunition. The kinetic energy APFSDS-T round is fired with a muzzle velocity of 5,900ft/s (1,800m/s) and has a maximum effective range of 2,300 yards (2,100m). The other anti-tank round is the shaped-charge type, which is officially known as HEAT-FS and which has a maximum effective range of 4,400 yards (4,000m). Finally, there is an HE-FRAG(FS) round, which would be used against targets such as bunkers, troops ▶

Above: Escorted by AFVs, Finnish Army T-72 MBTs press forward, their 2A46 125mm guns firing. The Finnish Army is just one of 14 export customers for the T-72, having acquired 80 examples. Iraqi Army T-72s were heavily used in the Gulf War, but their effectiveness was comprehensively neutralised by the Coalition anti-tank forces.

▶ in the open or light vehicles, and which has a maximum range in the indirect fire mode of 10,300 yards (9,400m).

The tank normally carries 39 rounds: 12 APFSDS-T, 21 HE-FRAG-FS and 6 HEAT-FS. All three types of ammunition come in two parts: the projectile and the cartridge case, the latter being entirely combustible apart from a small metal stub. The theoretical rate of fire is 8 rounds per minute, although whether this could be achieved, let alone sustained, on the battlefield is a different matter.

The suspension uses six large-diameter roadwheels mounted on torsion-bars. An unusual feature of the earlier production models is that the tracks are given a degree of protection by four spring-loaded skirt plates, which spring forward at an angle of about 60° when unclipped prior to going into action. This supposedly gives protection against HEAT rounds, but seems of dubious value and has not been repeated on later versions.

There is a single, narrow snorkel tube for deep-fording rivers, which fits over the gunner's periscope mounting. Such a tube provides no means of escape for the crew in the event of something going wrong whilst underwater and the whole tactic of fording is very unpopular with the soldiers serving in the Soviet armoured units.

There have been numerous versions of this MBT since the basic T-72 first appeared. There have also been many minor variations within the major types, leading to confusion in the designations.

The original version, produced only in limited numbers, was the T-72A, armed with the same 2A26 gun as the T-64, most of which were subsequently rebuilt to have plastic armoured skirts and laser rangefinders. The initial major production model was the T-72B which was the first to have the definitive 2A46 gun and other improvements; most have later been given extra armour and a laser rangefinder.

Soviet production switched to an even more advanced model in the late 1970s, known as the T-72M, which has appeared in a number of versions. The original T-72M had many improvements over the T-72B, including additional armour on the turret, plastic laminate side skirts and a new laser rangefinder. Next came the T-72M1, with a large sheet of appliqué armour on the glacis plate, while later versions of this model also had a new type of armoured matting applied to the turret roof and either side of the driver's station. All this additional armour on the tank's upper surfaces was a clear response to the threat posed by NATO's "top-attack" munitions, which were starting to enter service at this time.

The T-72M2, which is also known by the NATO designation Soviet Medium Tank M1986 is a development of the T-72M1, with a turret which not only has thickened side armour, but also has an additional sheet of appliqué armour on the roof. A later version of the T-72M2 has explosive reactive armour (ERA) blocks added on the glacis plate (one layer), around the turret (one to three layers) and along the sides of the tank (one layer) to offer some degree of protection for the suspension and upper hull. All of this beefing up of the T-72's armour, particularly the upper armour, has resulted in a somewhat derogatory nickname — the Super Dolly Parton Tank!

The T-72M2 also sports a new smoke generator configuration, with a total of eight mortars being mounted on the left-hand side of the gunner's hatch. An ammunition box for the 12.7mm machine-gun has been deleted, and targeting at night enhanced by virtue of an enlarged night-sight housing for the gunner.

The latest to appear (in 1987) is the T-72MS which appears to be a major redesign of the whole vehicle. It has a new suspension, a new engine and ERA. It is also the first version of the T-72 capable of launching the AT-8 "Songster" anti-tank guided missile.

There are a variety of special versions of the T-72. The T-72K is a command tank, with extra communications, for use at battalion and regimental headquarters. The BREM-1 is an armoured recovery and repair vehicle and the IMR-2 is a combat engineer vehicle. The T-72's operational versatility has been further enhanced by virtue of its ability to carry mine-clearing equipment. With the exception of those tanks configured for command duties, all models of the T-72 can carry such equipment at the front of the hull. Polish Army T-72s have been noted with an alternative mine-clearing system in place, this being the PW-LWD rocket-

Above: An excellent study of T-72s on the move, in this case six of the 400 examples acquired by the Polish Army (note the national insignia diamond displayed prominently on each turret). In each case, the tank commander is manning the NSVT 12.7mm machine-gun. Polish T-72s have been seen carrying the PW-LWD mine-clearing system, this being positioned on the hull top towards the rear.

propelled explosive-filled chord carried in a container mounted atop the hull towards the rear of the tank.

The T-72 is in use with at least 15 armies and many thousands have been produced in four State armaments factories in the Soviet Union. The T-72G is in production in Czechoslovakia and Poland, both of which simply designate it the T-72. These are little different to the Soviet versions, but the Romanian version, designated TR-125, has a number of differences which raise the combat weight to some 47.27tons (48tonnes). The T-72 M1 is also in production in India and the T-72G in Iraq (where it is called the *Assad Babyle* or Lion or Babylon). Finally, it is in production in Yugoslavia, where it is designated M-84. This differs in a number of ways from the Soviet T-72, mainly in the optics and electronics. A number of Yugoslav M-84s have also been exported to Kuwait.

T-80 Main Battle Tank

Country of origin: Soviet Union.
Crew: 3.
Armament: One 2A46 125mm gun; one PKT 7.62mm co-axial machine-gun, one NSVT 12.7mm anti-aircraft machine-gun.
Armour: Classified.
Dimensions: Length (including main armament) 32ft 6in (9.9m); length (hull) 24ft 3in (7.40m); width 11ft 2in (3.40m); height 7ft 3in (2.20m).
Weight: Combat 94,715lb (43,000kg).
Ground pressure: 11.80lb/in² (0.83kg/cm²).
Engine: Gas turbine engine developing 985hp.
Performance: Road speed 47mph (75km/h); road range 250 miles (400km); vertical obstacle 2ft 11in (0.90m); trench 9ft 6in (2.9m); gradient 60 per cent.
History: T-80 entered service with the Soviet Army in 1985.
(*Data apply to gas turbine-powered T-80*).

Design and development of the T-80 was undertaken by the AF Kartsev design bureau, located in the Ural fighting vehicle factory at Nizhni-Tagil, which was also responsible for the T-64 MBT. Development started in the mid-1970s, with production starting around 1983 and service introduction around 1985.

The T-80 incorporates numerous advances over the T-64, of which the most significant is the installation of a gas turbine engine, possibly influenced by the use of a similar powerplant in the United States Army's M1 Abrams MBT. Little is known about the Soviet gas turbine, except that it develops approximately 985hp and is coupled to a manual transmission with five forward and one reverse gears. Two additional jettisonable fuel tanks are mounted on the rear of the T-80 and a third tank can be fitted on the engine covers. It is significant that the latest version of the T-80 (described in more detail below) is powered by a diesel engine, so it must be presumed that the Soviet Army was not satisfied with either the overall performance of the gas turbine or its high fuel consumption — or, possibly, both!

The hull of the T-80 is of steel, with laminated armour in various crucial areas such as the glacis plate. The turret is of cast steel, but has an inner layer of "special" armour. The inside of the driver's compartment and the turret are lined with a special synthetic, lead-based material, similar to that on the T-72, which provides protection against neutron radiation and electro-magnetic pulses.

The suspension uses six road wheels on each side, mounted on torsion bars. This is a reversion to the more traditional Soviet system, compared to the T-64's small wheels and hydropneumatic suspension.

The main weapon is the 2A46 125mm smoothbore gun with a horizontal loader, which appears, at least from external inspection, to be identical to that used in

the T-72. Some of the T-80 fleet are equipped to fire the AT-8 "Songster" anti-tank guided-weapon, although it is not certain that all are. The gun fires the same rounds as that mounted in the T-72: two types of APFSDS-T, HEAT-FS and HE-FRAG(FS).

As with all Soviet tanks the T-80 is fitted for deep wading. A large cylindrical container is mounted across the rear of the turret, which carries two snorkel tubes. One fits over the gunner's periscope mounting, the other over the radiator grill to provide an air-intake for the gas turbine.

A new version of the T-80 was revealed in 1987, although it had already been in service for some years. Generally known by its US Army designation of T-80 Model 1984, it has between 185 and 220 explosive reactive armour (ERA) blocks mounted on the glacis plate and turret. These ERA blocks explode when hit by an incoming HEAT round, initiating the high-powered, molten metal jet, which then dissipates its energy in penetrating the block, leaving insufficient energy to then penetrate the main armour of the tank. Such devices provide very significant protection against HEAT rounds fired from guns and missiles fitted with HEAT warheads.

Yet a further improved version of the T-80 appeared in 1989, whose only known designation is Soviet Medium Tank M1989, or SMT 1989. In this version the gas turbine is replaced by a diesel engine, presumably a similar model to the V-12, 785hp unit fitted to the T-72. The most obvious external change is a new type of additional armour protection on the turret, in which a metal cowl has been fitted over the bank of ERA blocks, resulting in a marked change to the tank's appearance. The ERA blocks on the glacis plate have also been rearranged to give a neater and more comprehensive coverage. There are many other changes, including remote-control firing for the roof-mounted 12.7mm anti-aircraft machine-gun, additional smoke grenade dischargers and an improved on-board fire control system.

Above: The latest Soviet MBT is the T-80, a tank which incorporates features from both the T-64 and the T-72 and which is armed with a 125mm gun firing separate loading ammunition. Four types of shell are fired.

Left: The T-80, like the T-64, has been fitted with explosive reactive armour to give a massive increase in protection against the latest generation of anti-tank guided weapons, which carry HEAT warheads.

155mm Bandkanon 1A Self-propelled Gun

Country of origin: Sweden.
Crew: 5.
Armament: One 155mm gun; one 7.62mm anti-aircraft machine-gun.
Armour: 0.79in (20mm).
Dimensions: Length (overall) 36ft 1in (11m); length (hull) 21ft 6in (6.55m); width 11ft 1in (3.37m); height (with anti-aircraft machine-gun) 12ft 8in (3.85m).
Weight: Combat 116,850lb (53,000kg).
Ground pressure: 12lb/in² (0.85kg/cm²).
Engines: One Rolls-Royce K.60 diesel developing 240hp at 3,750rpm, and one Boeing Model 502/10MA gas turbine developing 300shp at 38,000rpm.
Performance: Road speed 17mph (28km/h); range 143 miles (230km); vertical obstacle 3ft 2in (0.95m); trench 6ft 7in (2.0m); gradient 60 per cent.
History: In production 1966-68. 30 in service with Swedish Army.

The Bandkanon 1A, or VK-155 as it is also known, is one of the heaviest self-propelled guns in service anywhere in the world. The prototype was built by the famous Bofors Ordnance Company in 1960; but the type was not produced in large numbers, staying in production for only two years. The VK-155 shares many automotive components with the S-tank, for example the power pack, which was also designed and built by Bofors. The driver is seated in the front part of the hull whilst the other four crew members are seated in the large turret at the rear of the hull. The 155mm gun has an elevation of + 40° and a depression of − 3°, and traverse is 15° left and 15° right. Elevation and traverse are both

continued ▶

Above: The 155mm gun of the Bandkanon 1A has a maximum elevation of +40° and a depression of −3°, and is mounted in a fully armoured turret which can be traversed 15° left and 15° right.

Left: Many of the automotive components of the 155mm Bandkanon 1A are identical to those of the S-tank which was also designed and manufactured by the famous Bofors company for the Swedish Army.

powered, but manual controls are provided for use in an emergency. The gun is fed from a magazine which holds 14 rounds in two layers of seven rounds, allowing the weapon to achieve a high rate of fire — a complete magazine in one minute. Once the magazine is empty a full magazine is brought up by a truck and loaded in place of the empty magazine, which takes about two minutes. The 155mm gun fires its HE round to a maximum range of 23,410 yards (21,400m). As soon as the weapon has fired the required number of rounds it would normally move to a new fire position before the enemy could pinpoint its exact position and return fire. A 7.62mm machine-gun is mounted on the left side of the turret, and can be used against both ground and air targets. The chassis has six road wheels, with the drive sprocket at the front. The suspension, which is of the hydro-pneumatic type, is locked in position when the gun is fired, thus providing a more stable firing platform. Although a unique gun, the VK-155 has a number of drawbacks. It is very heavy, rather slow and difficult to move across some bridges and roads. It is not possible to change types of ammunition quickly. For

example, a forward observer may ask for five rounds of HE on a target, followed by smoke rounds. But unless the magazine has a smoke round at that time it would not be able to comply. Moreover, unlike most other Swedish AFVs, for example the S-Tank, the Pbv.302 APC and the Ikv.91 tank destroyer, the VK-155 has no amphibious capability at all. Apart from the 30 Bk 1As, the Swedish Army now employs only towed weapons, mainly the Bofors-designed FH-77A and FH-77B 155mm towed howitzers.

Below: A 155mm Bandkanon 1A self-propelled gun in the firing position, and showing the magazine located at the rear which holds up to 14 high-explosive (HE) rounds of ammunition in two layers of seven rounds apiece. Also protruding to the rear is the ammunition handling system, while at the other end, the 155mm gun is fitted with a pepperpot muzzle brake. The gun's maximum elevation is +40°.

Infanterikanonvagn 91
Light Tank/Tank Destroyer

Country of origin: Sweden.
Crew: 4.
Armament: One 90mm gun; one 7.62mm machine-gun co-axial with main armament; one 7.62mm anti-aircraft machine-gun; 12 smoke dischargers.
Armour: Classified.
Dimensions: Length (with gun forwards) 29ft (8.85m); length (hull) 21ft 0in (6.41m); width 9ft 10in (3m); height 7ft 7in (2.32m).
Weight: Combat 35,934lb (16,000kg).
Ground pressure: 6.97lb/in² (0.49kg/cm²).
Engine: Vovlo-Penta TD 120A 6-cylinder turbo-charged diesel developing 330hp at 2,200rpm.
Performance: Road speed 40mph (65km/h); range 310 miles (500km); vertical obstacle 2ft 8in (0.8m); trench 9ft 2in (2.8m); gradient 60 per cent.
History: Entered service with Swedish Army in 1965. Production complete.

The Swedish Army is required to be able to fight a potential enemy in the forests and among the many lakes of its native country, in temperatures which can be as low as −35°C (−31°F). Further, as a neutral country, it must be prepared to fight any enemy and without the aid of allies. It is thus not surprising that, faced with these unusual requirements, the Swedish Army should produce some unusual military equipment.

The Ikv-91 is the outcome of a contract placed with Hägglund Vehicle AB in 1968 for a new combat vehicle to replace the Strv 74 light tank, Ikv-102 and Ikv-103 infantry cannon, and the Pansarvarnskanonvagn m/63 then in service. The result was the Ikv-91, which is a cross between a light tank and a self-propelled anti-tank gun. Some 200 have been produced for the Swedish Army, production having started in 1975 and been completed in 1978. The Ikv-91 and a new version with a 105mm gun (Ikv-105) have both been offered for export, but no orders have been received.

▶

Above: The Ikv 91 fires its L/54 90mm gun from cover. The low-pressure gun is claimed to create less recoil loads and reduce muzzle effects (flash, smoke, thrown-up dust) than others.

Below: Sweden's Ikv 91, which is designed to operate with anti-tank units in almost any terrain. It has good cross-country performance, and operates well in regions where there is marshy ground and water obstacles, such as rivers and lakes.

Above: In water the lkv 91 is propelled by its tracks giving it a speed of 4.36mph (7km/h), sufficient to cross moderately fast-flowing rivers. For amphibious operations a trim vane is erected and low screens are raised for air inlets, outlets and exhaust.

Below: The lkv 91's low silhouette, well profiled glacis plate and turret front are intended to afford protection, although priority has been given to high mobility.

The hull of the Ikv-91 is of all-welded steel construction and is divided into three compartments: driver's at the front, fighting compartment in the centre and the engine at the rear. The driver is seated on the left at the front of the tank, with the other three crew members seated in the all-welded turret, with the commander and gunner on the right and loader on the left. The main armament is a Bofors-designed low-pressure gun, firing fin-stabilised, high-explosive and high-explosive anti-tank rounds. 59 rounds are carried, 18 of which are stowed forward, beside the driver. Elevation limits for the gun are +15°/−10°. The barrel is fitted with a fume extractor and has recently been retrofitted with a thermal sleeve. The 360° turret traverse is powered by an electro-hydraulic system, with manual reversion. The gunner's optical sight incorporates a laser rangefinder, which gives a high probability of a first-round hit. There is a co-axial 7.62mm machine-gun, with a second 7.62mm anti-aircraft machine-gun on a flexible mount above the loader's hatch. There are also 12 smoke-grenade dischargers, six on either side of the turret.

The engine is a Volvo-Penta four-stroke, 6-cylinder diesel, which produces 360hp at 2,200rpm; it is mounted diagonally to conserve space. The fully automatic transmission system is provided by Allison and gives four forward and one reverse gear. To cope with starting conditions in the severe winter conditions met in the north of the country there is a built-in blow-torch to pre-heat the engine, a feature not found on many tanks!

The torsion-bar suspension utilises six large, rubber-tyred road wheels and the track has been specially designed by Hägglund for use in snow conditions; there are no return rollers. Studs or 50mm long conical spikes can be fitted to enhance the performance in deep snow — a valuable asset given the harsh Scandinavian climate.

There are many lakes in Sweden and it was thus essential that the Ikv-91 should be fully amphibious, which is part of the reason for its somewhat bulky appearance. There is a trim vane on the glacis plate which is erected before entering the water and low screens are raised around the air inlets and the exhausts. When swimming, the Ikv-91 is propelled by its tracks, which give a maximum speed of 4.3mph (7km(h).

A new version appeared in the 1980s, armed with a Bofors 105mm low recoil gun, which can fire all the internationally standard range of 105mm ammunition. Designated the Ikv-91-105 this vehicle has been tested by various countries, including India, but no orders have been placed.

Stridsvagn Strv-103
Main Battle Tank

Country of origin: Sweden.
Crew: 3.
Armament: One L74 105mm rifled gun; two 7.62mm co-axial machine-guns, one 7.62mm anti-aircraft machine-gun (commander); eight smoke dischargers (four on each side of turret).
Armour: All-welded steel.
Dimensions: Length (including main armament) 29ft 6in (8.99m); length (hull) 23ft 1in (7.04m); width 11ft 11in (3.63m); height 7ft 11in (2.43m).
Weight: Combat 112,335lb (51,000kg).
Ground pressure: 14.80lb/in² (1.04kg/cm²).
Engine: Detroit Diesel 6V-53T 6-cylinder, water-cooled, diesel engine developing 290hp at 2,800rpm.
Performance: Road speed 31mph (50km/h); range 242 miles (390km); vertical obstacle 2ft 11in (0.9m); trench 7ft 6in (2.3m); gradient 60 per cent.
History: Entered service with the Swedish Army in 1966. 300 in service.

Above: Although never tested in battle, the highly unconventional nature of the S-tank's design and appearance led to a general re-appraisal of MBT design philosophies. In reality, however, it has proved somewhat ineffective.

Right: A graphic illustration of the Strv-103C's peculiar hydropneumatic suspension, at its maximum elevation of +12°. This side profile also highlights the unique outline of this Swedish design, due to the singular lack of any conventional turret.

At the time it entered service, the Swedish Stridsvagn Strv-103 (popularly known as the "S-tank") caused great interest as it appeared to provide a new way forward for tank design, leading to lighter and more mobile three-man MBTs. Its design originated in the 1950s, with the Swedish firm of Bofors being awarded the development contract in 1958. The first two prototypes were completed in 1961, followed by 300 production tanks produced between 1966 and 1971.

The S-tank's crew are all located in a central fighting compartment. The driver/gunner is on the left, facing forwards; behind him sits the radio operator, who faces to the rear and who drives the tank backwards, when required. The commander sits on the right of the gun and also has an accelerator and brake to control the vehicle, if required.

Main armament is the L74 105mm rifled gun, a lengthened version of the British L7 gun, produced in Sweden. The barrel is mounted rigidly in the glacis plate thus doing away with the need for a turret. This results in reduced overall height for the tank and reduced weight, since there is no turret or recoil mechanism, and also enables an automatic loader to be installed. This device holds 50 rounds, a mix, as required by the tactical situation of APDS, HESH, HE and smoke. The tank can fire 10-15 rounds per minute and empty cases are automatically ejected through a hatch in the rear of the hull.

The power pack in the S-103A and S-103B versions was a Rolls-Royce K60 diesel, which was used for normal operations, and a Boeing gas-turbine which was brought in to provide additional power for combat or when crossing difficult terrain. In the Strv-103C, however, the Rolls-Royce engine has been replaced by a more powerful Detroit Diesel 6V-53T, although the gas-turbine has been retained unchanged.

The hydropneumatic suspension is used to aim the gun. The gun is laid in elevation by the driver/gunner, who adjusts the suspension to alter the elevation between + 12° and − 10°. The gun is traversed by slewing the tank in its tracks. When the gun is fired the suspension is locked to provide a stable platform.

The S-tank bristles with innovations, which caused much excitement when it first appeared. It was widely tested, the British Army even leasing sufficient to equip a complete armoured squadron in Germany for a protracted field trial. However it has proved to be less successful than first thought. It must expose a large cross-sectional area when in a hull-down firing position and cannot fire with any accuracy on the move. No further designs of this type have been produced.

The next-generation Swedish MBT is already under development. After examining various radical designs, such as articulated vehicles, the Swedish General Staff appears to have settled on an essentially conventional design. Designated the *Stridsvagn 2000*, this new MBT will weigh some 57 tons (58 tonnes) and will be armed with a new 140mm gun in a conventional traversing turret, together with a coaxial 40mm Bofors cannon.

Pz 61 & 68 Main Battle Tanks

Country of origin: Switzerland.
Crew: 4.
Armament: One L7 105mm rifled gun; one 7.5mm machine-gun co-axial with main armament; one 7.5mm anti-aircraft machine-gun, three smoke dischargers on each side of turret.
Armour: 2.36in (60mm) maximum.
Dimensions: Length (including main armament) 31ft 1½in (9.49m); length (hull only) 22ft 8in (6.9m); width 10ft 3½in (3.14m); height (overall) 9ft (2.75m).
Weight: Combat 87,523lb (39,700kg).
Ground pressure: 12.23lb/in² (0.86kg/cm²).
Engine: MTU Mb 837 eight-cylinder diesel developing 704hp at 2,200rpm.
Performance: Maximum road speed 34mph (55km/h); range 217 miles (350km); vertical obstacle 2ft 6in (0.75m); trench 8ft 6in (2.60m); gradient 60 per cent.
History: Production complete. In service with the Swiss Army (390).

In the early 1950s design work started on a Swiss MBT, with the first prototype, the Pz 58, being completed in 1958. This eventually entered production as the Pz 61, a conventional 83,702-lb (38,000kg) MBT which was armed with a Swiss-produced version of the British L7 gun and a co-axial Oerlikon 20mm cannon.

The design was improved in the light of experience, which led to the Pz 68, of which 170 were built. This MBT has an improved fire control system and the main armament is stablised in both horizontal and vertical planes. The engine is also more powerful than that of the Pz 58 and a more efficient gearbox is fitted.

The driver sits centrally in the front of the hull. The other three crew members are in the turret, with the commander and gunner on the right of the main gun and the loader on the left. The MTU engine is imported from Germany, but the transmission and gearbox are Swiss.

The main armament, an L7 105mm rifled gun, has an elevation of + 21° and can be depressed to − 10°. There is a 7.5mm machine-gun mounted co-axially with the main gun and a similar weapon is mounted above the loader's hatch for use in the anti-aircraft role.

170 of the Pz 68 Series 1, MBTs were delivered between 1971 and 1974. These were followed by 50 Pz 68 Series 2 MBTs which have an alternator, a thermal sleeve for the gun and a fume scavenging system for the fighting compartment. Next to appear was the Series 3 (110 built in 1978-79) and the virtually identical Series 4 (60 delivered in 1983-84) which differed from the Series 2 only in having a larger turret. In a recent programme, 195 Pz 68s will be upgraded by fitting an improved fire control system and by replacing the fire suppression system in the fighting and engine compartments. The suspension will also be upgraded, a muzzle reference system fitted and GRP fuel tanks added.

Above: An Entp Pz 65 armoured recovery vehicle changes the power pack of a second Entp Pz 65. The 'A'-frame mounted at the front of the hull lifts a load of 33,040lb (15,000kg), and the vehicle has two winches and one dozer blade.

Left: The Pz 61 is armed with an L7 105mm gun designed in the UK and made under licence in Switzerland. Unlike most other tanks, the single 7.62mm machine-gun is used by the loader rather than by the tank commander.

AS-90 155mm
Self-propelled Gun

Country of origin: United Kingdom.
Crew: 5.
Armament: One 155mm gun; one 12.7mm anti-aircraft machine-gun.
Dimensions: Length (gun forward) 31ft 10in (9.71m); length (chassis) 22ft 11in (6.99m); width 10ft 10in (3.30m); height 9ft 10in (3.0m).
Weight: 92,593lb (42,000kg).
Engine: Cummins VTA 903T-660 V-8 diesel developing 660hp at 2,800rpm.
Performance: Road speed 34mph (55km/h); road range 267 miles (430km); vertical obstacle 2ft 11in (0.88m); trench 9ft 2in (2.80m); gradient 60 per cent.
History: Entered service with the British Army in 1992.

In the 1960s the British and German armies, later joined by the Italians, conducted a very successful collaborative project which resulted in the towed 155mm Field Howitzer-70 (FH-70). This led to a second project to develop a self-propelled (SP) 155mm howitzer, designated the SP-70, which was tri-national from the start. This started in 1973, but unfortunately it proved to be too ambitious and after many years of work and a great deal of expense the UK withdrew from the project in 1986.

This left the UK in the embarrassing position of having no new successor for the ageing Abbot, which was the only 105mm SP left in front-line service with a major NATO army. It would obviously have been possible to buy more of the United States' 155mm M109 SPs, which were already in service with the British Royal Artillery in some numbers, but this, too, was an elderly design. Fortunately ▶

Below: The British Army has so far ordered 179 AS-90 155mm SPs as replacements for the elderly Abbot 105mm SP.

▶ the British armaments firm of Vickers Shipbuilding and Engineering Ltd (VSEL) had forseen this possible gap in the market and had been quietly developing a totally new 155mm SP howitzer, designated the Artillery System-90 (AS-90), which marries the 39-calibre 155mm ordnance of the FH-70 with a totally new recoil system and tracked chassis. After an assessment of various bids, the British Army selected the AS-90 in June 1989, with an initial order for 179. These will equip six artillery regiments, each with 24 weapons in three eight-gun batteries.

The AS-90 chassis is of all-welded, steel construction. The driver sits at the front on the left, with the engine to his right. The engine is an American-designed Cummins diesel, which is a developed and more powerful version of that fitted in the US Army's M2/M3 Bradley Infantry Fighting Vehicle. The transmission is the Zahnradfabrick LSG-2000 with four forward and two reverse gears.

The fighting compartment is at the rear, with a large, all-welded, steel turret housing the four-man gun crew. The turret ring is no less than 8ft 10in (2.70m) in diameter and is fairly high, giving a particularly good working environment. Inside the turret the commander and layer sit on the right, with the two ammunition numbers on the left.

The suspension system uses six road wheels, each with fully-independent hydrogas suspension. This has a number of significant advantages. First, the advanced recoil system coupled with the hydrogas suspension mean that there

Below right: A full 360° traverse has been built into the AS-90's four-man turret. Elevation limits are +70° to −5°.

Below: A look at the "business end" of the AS-90 reveals the 155mm gun's double baffle muzzle-break and fume extractor.

is no requirement for a spade, thus saving weight and complexity. Secondly, the absence of torsion bars mean that there is no need for a false floor for the turret, thus contributing to the good headroom in the turret. Thirdly, the suspension gives the vehicle an excellent cross-country ride, enabling a high average speed to be maintained.

The barrel is the Royal Ordnance 39-calibre 155mm, fitted with a double baffle muzzle-brake and a fume extractor. Elevation limits are $+70°/-5°$, there is full 360° traverse and the gun can be reloaded at any angle. Maximum range with a standard NATO 155mm HE round is 27,000 yards (24,705km) and 35,000 yards (32,025m) with an extended-range round. 48 projectiles are carried and are transferred from their storage bins to the breech by a semi-automatic system, with manual operation, if required. A burst rate of three rounds in 10 seconds is possible, with an intense rate of six rounds per minute for three minutes. Normal sustained rate is two rounds per minute for one hour, although using modern "shoot-and-scoot" tactics to avoid enemy counter-battery fire it is highly improbable that an AS90 would ever remain in one position for such a long time.

The vehicle and turret have been designed to include development potential. VSEL state that the new NATO standard 59-calibre barrel can be accommodated without any problems of space or stability, while there are also plans to use the chassis as the basis of a family of armoured fighting vehicles.

Centurion Main Battle Tank

Country of origin: United Kingdom.
Crew: 4.
Armament: One L7 series 105mm gun; one .3in machine-gun co-axial with main armament; one .5in ranging machine-gun; one .3in machine-gun on commander's cupola; six smoke dischargers on each side of the turret.
Armour: 0.67in-6.08in (17mm-152mm) maximum.
Dimensions: Length (gun forward) 32ft 4in (9.85m); length (hull) 25ft 8in (7.82m); width (including skirts) 11ft 1½in (3.39m); height 9ft 10½in (3.01m).
Weight: Combat 114,250lb (51,820kg).
Ground pressure: 13.5lb/in² (0.95kg/cm²).
Engine: Rolls-Royce Meteor Mk IVB 12-cylinder liquid-cooled petrol engine developing 650bhp at 2,550rpm.
Performance: Road speed 21.5mph (34.6km/h); range 118 miles (190km); vertical obstacle 3ft (0.91m); trench 11ft (3.35m); gradient 60 per cent.
History: Entered service with the British Army in 1949. Centurions remain in service with the armies of Denmark (226), Israel (1,080), Jordan (293), Kuwait (10), Singapore (63), Somalia (30), South Africa (300) and Sweden (350).

The Centurion tank was developed at the end of World War II and entered service with the British Army in 1949. 4,423 were produced, some 2,500 of them for

export. Those which entered service with the British Army have all been withdrawn and sold abroad; many other armies have scrapped their Centurions, but there are still some 2,000 or so in service around the world.

The Centurion has been a very successful, battleworthy and popular tank, and has also proved capable of accepting numerous improvements. The original model was armed with a 17-pounder (76.2mm) gun, which was later replaced by the 20-pounder (83.8mm) model; the great majority of those still in service, however, are armed with the British-designed L7 105mm gun. Most have also had more powerful diesel engines fitted to replace the original Rolls-Royce Meteor petrol engine (a modified version of the Rolls-Royce Merlin engine, which powered the Spitfire fighter) which was exceptionally thirsty and gave a very poor range.

Israel operates large numbers of Centurions which have been modified by fitting a Teledyne Continental AVDS-1790-2A diesel engine and an Allison CD-850-6 automatic gearbox. Designated the Upgraded Centurion, most of these MBTs are also fitted with IMI Blazer explosive reactive armour. The Israeli Army also uses the Centurion chassis as a basis for armoured recovery vehicles, armoured personnel carriers and 290mm multiple rocket launchers.

Jordanian Army Centurions have been given a similar upgrade and are also now powered by an Allison diesel engine; they are designated Tariq. South Africa, long cut-off from external arms supplies by the UN arms embargo, has upgraded its Centurions with a South African-manufactured L7A1 105mm gun and a V-12 diesel engine. The modified version is known as the Olifant.

Sweden continues to upgrade its Centurions, which are known as the Strv

Above: As with most MBT designs, the Centurion has lent itself well to adaptation, as illustrated in this view of an Armoured Engineering Vehicle. Atop the glacis plate is a length of rolled aluminuim roadway, while behind the vehicle is a towed trailer.

Left: With over 1,000 examples still in service, the Israeli Army is by far the biggest user of the Centurion, most of which have been fitted with the Teledyne Continental AVDS-1790-2A diesel engine and an Allison CD-850-6 automatic gearbox.

101 and 102. Allison diesels and gearboxes have been installed and new sights, fire control systems and suspension fitted. There is even talk of fitting a 120mm gun to increase its "punch". The Austrian Army uses dismounted Centurion turrets in the static defence role.

Challenger Main Battle Tank

Country of origin: United Kingdom.
Crew: 4.
Armament: One Royal Ordnance L30 120mm rifled gun; one McDonnell Douglas 7.62mm co-axial Chain-Gun; one L37A2 7.62mm anti-aircraft general-purpose machine-gun; two five-barrel smoke grenade dischargers.
Armour: Chobham.
Dimensions: Length (including main armament) 37ft 11in (11.55m); length (gun to rear) 32ft 4in (9.86m); width 11ft 7in (3.52m); height 8ft 2in (2.49m).
Weight: Combat 127,775lb (60,000kg).
Ground pressure: 12.8lb/in² (0.9kg/cm²).
Engine: Perkins CV12 TCA 12-cylinder 60°V direct injection 4-stroke liquid-cooled diesel engine developing 1,200bhp at 2,300rpm. Perkins 4.108 4-stroke diesel auxiliary power unit.
Performance: Road speed 35mph (56km/h); range 340 miles (550km); vertical obstacle 2ft 11in (0.9m); trench 7ft 8in (2.34m), gradient 60 per cent.
History: Entered production in 1990; will enter service with British Army in 1992.

With the Chieftain accepted for service in 1963, the thoughts of the British Army turned towards the next generation tank. A national project was started in the late 1960s and this work continued, albeit at a low tempo, during the period of the Anglo-German MBT project. This started in 1970, but, as with every other collaborative MBT project, the partnership ended in 1977, whereupon attention reverted to the national work. Meanwhile, the Chieftain had attracted orders from ▶

Above: A sight enemy ground forces dread — a Challenger 1 MBT smashing through a defensive sand berm on its way to battle.

Below: The well-sloped front of the Challenger's low profile turret is a deliberate design attempt to decrease the chances of an anti-tank missile team achieving a successful "kill".

Below: While the obvious purpose of a Main Battle Tank is to fire the shells carried within, the ability to use external space for some improvised logistical self-support is known to every good tank crew. Photographed during Operation *Desert Storm*, this Challenger 1 proves the point, with a pair of auxiliary fuel drums attached to the rear of the hull and a group of "top up" fuel cannisters strapped down to the main body immediately behind the turret.

▶ the Shah of Iran and a developed version had been produced designated the Shir 1. This led in turn to the Shir 2, a very much better MBT, whose development was mainly funded by the Iranian order. Unfortunately the Iranian order then fell through, due to the revolution which ousted the Shah, and Vickers then produced a new version of the Shir 2, altered to suit British Army requirements. This was ordered into production in 1978 as the Challenger 1, entering service with the British Army in 1984, with a total of 420 having been delivered by the time production ended in mid-1990.

There was considerable bad publicity for Challenger 1, which reached its peak when the British Army decided to withdraw from the Canadian Army Trophy, an international tank gunnery contest among NATO teams in the Northern and Central Army Groups. Intense speculation about a future MBT for the British Army culminated in the late 1980s with an announcement that there would be an international competition for what was officially termed the "Chieftain Replacement Programme". The four competitors were the Leopard 2 (Improved) from Germany, the M1A1 Abrams from the USA, the Leclerc from France and the Challenger 2 from the UK. After an intensive evaluation of the competing designs the Challenger 2 was selected, subject to passing a series of "milestones" which would demonstrate that it fully met the British Army's requirement.

Meanwhile, the Gulf War broke out and the British Army sent an armoured division to Saudi Arabia, which included 176 Challenger 1s. These performed exceptionally well, showing a high degree of mechanical reliability, the tanks in the forward armoured regiments covering an average of 217 miles (350km) each in the 100-hour ground war, and there were just two breakdowns in the entire force. The 120mm gun proved very accurate and more than half the engagements involved the use of High-Explosive Squash Head (HESH) rounds. Following this great success and the passing of the milestones, all of which was accompanied by some intense political lobbying, an order was finally placed for Challenger 2 MBTs.

The Challenger 2's hull is similar to that of the Challenger 1. The driver sits centrally, with an unusual trough in the glacis plate to enable him to see. The other three crewmen are in the turret, with the commander and gunner on the right of the gun, and the loader on the left. The hull and turret are constructed of welded steel and Chobham armour, and are of an exceptionally good ballistic shape.

The gun is the new L30 120mm rifled gun, which is fitted, as is normal British practice, with a thermal sleeve, fume extractor and muzzle reference system. This new gun is part of the Challenger ARMament package which includes the gun, a new charge system, and a new and more effective anti-tank round with a depleted uranium (DU) penetrator. Sixty four projectiles are carried, together

with 42 charges, the latter being stowed in armoured boxes below the turret ring for maximum safety. The L30 gun is also being retrofitted to the Challenger 1 tanks, replacing their L11A5 guns.

The Challenger 2 is now the only new MBT of the 1990s to mount a rifled 120mm main gun. This is due to the British Army's continuing belief in the value of the HESH round. When it hits a tank the high-explosive on this type of round forms, for the briefest moment of time, a circular "cake" which is then exploded by a charge at the base of the projectile. The shock from this explosion dislodges a large scab from the inside wall of the tank which then ricochets at high velocity around the crew compartment. The British Army firmly believes that this type of round is needed to complement the high-velocity kinetic energy round (APDS, APFSDS, etc.) and has therefore insisted on retaining a rifled barrel, since such a round depends on spin-stabilization for in-flight stability and cannot be fired from a smoothbore barrel.

A McDonnell Douglas Helicopters 7.62mm Chain-Gun is mounted coaxially with the main gun and a second 7.62mm anti-aircraft machine-gun is mounted on the turret roof.

The power unit is a Perkins diesel with a rated output of 1,200hp. The TN37 transmission on the Challenger 1 has been described as being insufficiently flexible and is being replaced by the TN54 model, which has six forward and two reverse gears. This power train is, in fact, already being installed in the Challenger Armoured Repair and Recovery Vehicle (ARRV) and 12 of these were sent to Saudi Arabia for the Gulf War where they posted 100 per cent availability, a truly remarkable achievment.

Above: The sole variation on the Challenger MBT theme to date is the Armoured Recovery and Repair Vehicle, complete with dozer blade. A dozen of these vehicles were deployed to the Gulf in support of UK land forces, and proved their worth many times over in very arduous combat conditions.

Left: The British Army's MBT for the 1990s — Challenger 2. Toting an L30 120mm rifled gun, much is expected (both politically and militarily) of this design which won out over the M1A1 Abrams, Leopard 2 and Leclerc MBTs at the end of a hard-fought competition.

FV4201 Chieftain Main Battle Tank

Country of origin: United Kingdom.
Crew: 4.
Armament: One L11 series 120mm gun; one 7.62mm machine-gun co-axial with main armament; one 7.62mm machine-gun in commander's cupola; one .5in ranging machine-gun; six smoke dischargers on each side of turret.
Armour: Classified.
Dimensions: Length (gun forward) 35ft 5in (10.79m); length (hull) 24ft 8in (7.52m); width overall (including searchlight) 12ft (3.66m); height overall 9ft 6in (2.89m).
Weight: Combat 121,250lb (55,000kg).
Ground pressure: 14.22lb/in² (0.90kg/cm²).
Engine: Leyland L.60 No 4 Mk 8A 12-cylinder multi-fuel engine developing 750hp at 2,100rpm.
Performance: Road speed 30mph (48km/h); road range 280 miles (450km); vertical obstacle 3ft (0.91m); trench 10ft 4in (3.15m); gradient 60 per cent.
History: Entered service with the British Army in 1967. Now in service with the following armies: Iran (60-100); Iraq (150); Jordan (90); Kuwait (143); Oman (15); United Kingdom (870).

Following its experiences against German tanks in World War II the British Army has always given top priority to protection and firepower at the expense of mobility. Thus, when a requirement was issued in the 1950s for an MBT to replace the Centurion, the result was the best armoured tank of its generation, with the most powerful main gun. Seven prototypes of the new FV4201 Chieftain were completed between 1959 and 1962, and following protracted development problems with the engine, transmission and suspension, the first of some 900 Chieftains entered service with the British Army in 1967.

The Chieftain has a hull front of cast construction and the rest of the hull is welded; the turret is cast. The driver is placed centrally in the front of the hull in a semi-reclined position, which enables the overall height of the tank to be kept to a minimum. In the turret, the commander and gunner are on the right and the loader on the left. When originally introduced the gunner was provided with a 12.7mm ranging machine-gun, which fired rounds ballistically matched to the 120mm rounds, and was thus able to obtain precise ranging information for the main gun. This was simple and effective compared to the contemporary cross-turret optical rangefinders, but it was replaced by an even more effective method of range-taking in the 1970s, using the Barr & Stroud laser rangefinder. Later still, the Marconi Integrated Fire Control System was installed. Over 300 British Army Chieftains have also been fitted with the Thermal Observation and Gunnery Sight, which was developed for the Challenger MBT.

The L11A5 120mm rifled gun is fitted with a muzzle reference system, fume extractor and a thermal sleeve. The main rounds fired are Armour-Piercing Discarding Sabot, Armour-Piercing Fin-Stablised Discarding Sabot and High Explosive Squash Head.

Right: A Chieftain Mk 5 MBT with a thermal sleeve for its 120mm gun. Chieftains remaining in British Army service are now fitted with the GEC-Marconi Integrated Fire Control System which gives them an enhanced "kill" probability.

Above: A Chieftain MBT negotiating rough ground with relative ease.

Chieftain is fitted with a full range of night vision devices. All now have an infra-red (IR) detector, which can localise an IR source to within 62°. Various other devices have been developed for the tank, but are only fitted when the tactical situation requires it. These include a dozer blade and a deep fording kit.

In 1986, the British Army started a programme to fit appliqué armour, codenamed *Stillbrew*. This considerably enhances armour protection at comparatively small cost and with only a small effect on mobility. Other versions of the Chieftain include armoured repair and recovery vehicles, armoured recovery vehicles and the armoured vehicle-launched bridge. Part of the British Army's Chieftain fleet was replaced in the 1980s by the Challenger 1 and the remainder will be replaced in the 1990s by the Challenger 2.

The Iranian Army placed an order for 700 Chieftains in 1971, followed by another order for 125 of an improved version called the Shir 1 and 1,225 of an even more advanced version called the Shir 2. The Chieftains all arrived in Iran, but none of either version of the Shir were actually delivered, due to the collapse of the Shah's regime. Jordan subsequently ordered 90 Shir 1s, which they designated the Khalid, while the Shir 2 design became the basis of the British Army's Challenger 1 MBT.

Main armament of the Chieftain MBT consists of a 120mm rifled tank gun designed by the Royal Armament Research and Development Establishment at Fort Halstead. This fires a wide range of separate loading ammunition (eg, projectile and bagged charge) including Armour Piercing Discarding Sabot Tracer (APDS-T), High Explosive Squash Head (HESH) and Smoke, plus training rounds.

Vickers Main Battle Tank

Country of origin: United Kingdom.
Crew: 4.
Armament: One Royal Ordnance L7A1 105mm rifled gun; one 12.7mm co-axial ranging machine-gun; one co-axial L37A2 7.62mm general-purpose machine-gun; one L37A2 anti-aircraft machine-gun; two six-barrel smoke grenade dischargers.
Armour: Maximum 3.15in (80mm).
Dimensions: Length (including main armament) 32ft 1in (9.79m); length (hull) 24ft 10in (7.56m); width overall 10ft 5in (3.17m); height 10ft 1in (3.1m).
Weight: Combat 88,107lb (40,000kg).
Ground pressure: 12.66lb/in² (0.89kg/cm²).
Engine: Detroit Diesel 12V-71T turbocharged 12-cylinder diesel engine developing 720bhp at 2,500rpm. Perkins 4.10 4-stroke diesel auxiliary power unit.
Performance: Road speed 31mph (50km/h); range 330 miles (530km); vertical obstacle 2ft 9in (0.83m); trench 9ft 10in (3m); gradient 60 per cent.
History: Entered service with the Indian Army in 1965. Now in service: **Mark 1** — India (2,200), Kuwait (70); **Mark 3** — Kenya (76), Nigeria (72).

In the 1950s it was decided to set up a tank plant in India and teams were sent abroad to select a design which would meet the requirements of the Indian Army. The Vickers design was successful and in August 1961 a licensing contract was signed. Two prototypes were completed in 1963, one being retained by Vickers and the other being sent to India in 1964. Meanwhile plans were being drawn up for a factory to be built near Madras. Vickers delivered some complete tanks to India before the first Indian tank was completed early in 1969. These first tanks had many components from England, but over the years the Indian content of the tank steadily increased until the Indians built over 90 per cent of the tank themselves. Production eventually passed the two thousand mark, and the tank gave a good account of itself in the last Indian-Pakistani conflict. The Indians call the tank *Vijayanta* (Victorious). In designing the tank, Vickers sought to strike the best balance between armour, mobility and firepower within the limits of a tank weighing 85,045lb (38,610kg). The layout of the tank is conventional. The driver is seated at the front of the hull on the right with ammunition stowage to his left, and the other three crew members are located in the turret: the commander and gunner to the right and the loader to the left. The engine and transmission are at the rear of the hull. The engine and transmission are the same as those used in the Chieftain MBT. The suspension is of the torsion-bar type and consists of six road wheels with the drive sprocket at the rear and the idler at the front, there being three track-return rollers. The Vickers MBT is armed with the standard L7 series 105mm rifled tank gun, this having an elevation of + 20° and a depression of − 7°, traverse being 360°. A .3in machine-gun is mounted co-axially with the main armament and a similar weapon is mounted on the commander's cupola. Six smoke dischargers are mounted each side of the turret. Some 44 rounds of 105mm and 3,000 rounds of .3in machine gun ammunition are carried. The main armament is aimed with the aid of the ranging machine-gun method, which was used so successfully in the Centurion tank

Right. The Vickers Mark 3 MBT was built by India at Madras as part of a licence deal with the British designers. Over 2,000 examples of what the Indian Army knows as the *Vijayanta* (Victorious) eventually rolled off the production line. Main armament is a tried and tested L7A1 105mm rifled gun which was also manufactured in India under licence.

Above: Sporting an all-cast turret and a laser range-finder, the Vickers Mark 3 MBT has also found sales success in Kenya (76) and Nigeria (72). An enhanced version of the Mark 3, the Mark 3 (Improved), has yet to win an order.

Above: A Vickers Mark 1 MBT on the firing ranges. The 105mm gun in this model is aimed by a 12.7mm ranging machine-gun mounted co-axial with the main armament, firing in three-round bursts.

Above: A Vickers Mark 1 MBT powered by a General Motors 720bhp turbo-charged diesel engine, and sporting a thermal sleeve for its L7A1 105mm gun.

with the 105mm gun. The gunner lines up the gun with the target and fires a burst from the .5in ranging machine-gun, and can follow the burst as the rounds are all tracer. If they hit the target he knows the gun is correctly aimed and he can then fire the main armament. Some 600 rounds of ranging machine-gun ammunition are carried. Two types of main-calibre ammunition are used: High Explosive Squash Head and Armour-Piercing Discarding Sabot. A GEC-Marconi stabilisation system is fitted, and this enables the gun to be aimed and fired whilst the vehicle is moving. Indian production ended at 2,200. Kuwait also bought 70 Mark 1s, which were built by Vickers in the UK. A developed version, the Mark 3, was also built by Vickers, with 76 going to Kenya and 72 to Nigeria. In the mid-1980s Vickers carried out a redesign of the Mark 3, with an improved hull, better suspension, new powertrain, wider tracks and the latest electronic and optical devices. The result, the Vickers MBT Mark 3 (Improved), is a very modern-looking and capable tank, but has yet to win any production orders.

Vickers has also produced prototypes of the Advanced MBT Mark 7 in conjunction with Krauss-Maffei of Germany, which is basically a Leopard 2 chasis with the turret developed for another Vickers private venture MBT, the Valiant. Trials were successful, but no orders have been received.

Also in the mid-1980s, Vickers co-operated with FMC, a US company, to produce an export version of FMC's Close Combat Vehicle (Light) CCV(L). Relatively light and fast, the FMC 5 is armed with a 105mm gun and is protected by a mixture of aluminium and steel armour. No foreign orders have yet been placed for this interesting vehicle.

M1 Abrams Main Battle Tank

Country of origin: United States of America.
Crew: 4.
Armament: One M256 Rheinmetall 120mm smoothbore gun; one M240 7.62mm co-axial machine-gun; one M2 12.7mm and one M240 7.62mm anti-aircraft machine-gun. .
Armour: Classified.
Dimensions: Length (including main armament) 32ft 3in (9.83m); length (hull) 25ft 11in (7.92m); width 11ft 11in (3.66m); height 9ft 6in (2.89m).
Weight: Combat 125,890lb (57,154kg).
Ground pressure: 13.65lb/in² (0.96kg/cm²).
Engine: Lycoming Textron AGT-1500 gas-turbine developing 1,500hp at 3,000rpm.
Performance: Road speed 41mph (67km/h); range 300 miles (480km); vertical obstacle 3ft 6in (1.07m); trench 9ft 0in (2.74m), gradient 60 per cent.
History: First production M1 completed in 1980 and first production M1A1 in 1985. Egypt is to co-produce 555 M1A1s, starting in 1992.

In June 1973 contracts were awarded to both the Chrysler Corporation (which built the M60 series) and the Detroit Diesel Allison Division of the General Motors Corporation (which built the MBT-70) to build prototypes of a new tank designated M1, and later named the Abrams tank. These tanks were handed over to the US Army for trials in February 1976. In November 1976 it was announced after a four-month delay that the Chrysler tank would be placed in production. Production commenced at the Lima Army Modification Center at Lima in 1979 with the first batch of full production M1s being completed early in 1980.

The M1 has a hull and turret of British Chobham armour, which is claimed to make the tank immune to attack from both missiles and tank guns. Its crew consists of four, the driver at the front, the commander and gunner on the right of the turret, and the loader on the left. The main armament consists of a standard 105mm gun developed in Britain and produced under license in the United States and a 7.62mm machine-gun is mounted co-axially with the main armament. A 12.7mm machine-gun is mounted at the commander's station and a 7.62mm machine-gun at the loader's station. A total of 55 rounds of 105mm, 1,000 rounds of 12.7mm and 11,400 rounds of 7.62mm machine-gun ammunition are carried. The main armament can be aimed and fired on the move. The gunner first selects the target, then uses the laser rangefinder to get its range and depresses the firing switch. The computer makes the calculations and adjustments required to ensure a hit.

Right: The most recent — and by far the most important — addition to the US Army's MBT force during the 1980s, the baseline M1 Abrams has given rise to three distinct versions, with a fourth (the M1A2) now joining the ranks. In addition to its US Army service, the M1 Abrams has won orders from the USMC and the Egyptian Army.

Above: The angular front and sides of the M1's turret are among this MBT's most distinctive features. Atop the turret and to the left can be seen the M2 12.7mm AA machine gun.

Below: One of the prototype XM1 Abrams is put through its paces at the Aberdeen Proving Grounds in Maryland, USA. The first of 2,374 full production M1 Abrams MBTs were delivered to the US Army during 1980

▶ The M1 is powered by a multi-fuel Lycoming Textron gas-turbine. This has proved to be reliable in service and gives the M1 the high power-to-weight ratio of 27hp/tonne and the very rapid acceleration of 0-20mph (0-32km/h) in six seconds! It is also mechanically simple and particularly easy to service. Conversely, it is noisy and emits a very hot exhaust (and thus has a strong infar-red (IR) "signature"), but perhaps the most serious fault is that it is very thirsty on fuel. There is a fully-automatic transmission, equipped with four forward and two reverse gears.

After producing 2,374 basic M1s production switched in February 1985 to the M1 (Improved), which has better armour, but is otherwise identical to the basic M1. The major change came with the M1A1, which started to leave the production lines in 1987. This is armed with the Rheinmetall M256 120mm smoothbore gun, which had been originally developed in Germany to arm the Leopard 2. 40 120mm rounds are carried, compared to 55 rounds in the 105mm-armed M1. An integrated NBC protection system is also installed and the suspension system improved.

The most recent production version is the M1A1 with the Heavy Armor Package. In this, certain areas of the hull, particularly the front, are constructed of a new type of armour, consisting of depleted uranium (DU) encased in steel, which gives a density 250 per cent greater than that of normal steel. This is designed to counter the latest kinetic energy penetrators and will only be used for those M1A1s intended for deployment in Europe. The DU has a very low radiation emission rate, but raises the tank's overall weight.

Currently under development is a series of developments known as "Block II Improvements", which, if accepted, will result in the M1A2. The improvements involved are mostly associated with the command-and-control, electronic, optical and electrical systems, which, if implemented, will maintain the M1A1s position as the most sophisticated MBT in the world. ▶

Above: A tight squeeze for a US Army Europe M1 Abrams, as it trundles through a German town while on exercise.

Below: An impressive line-up of M1 Abrams in the deserts of Kuwait, awaiting the word to advance on Iraqi ground forces.

► 221 M1A1s are on order for the US Marine Corps to replace its ageing M60A1s. This model is almost identical to the Army's M1A1, except that all Marine Corps tanks will be fitted with the Deep Water Fording Kit for use in amphibious landings.

The M1A1 is to be produced in Egypt, under an agreement signed in 1988. The first tanks will be sent from the USA fully-assembled, to be followed by partly-assembled versions and then by kits. Egypt has already started the manufacture of certain components, but hulls, gun, ammunition and electronics will be supplied by the USA.

The delivery/order picture for the M1-series for the US forces at the end of 1991 was:

M1 (Basic)	2,374	— delivered		
M1 (Improved)	894	— delivered		
M1A1 — Army	4,199	— in production	**TOTAL**	7,750
USMC	221	— in production		
M1A2	62	— projected		

The picture for the future is somewhat uncertain as a result of the reduction of tension resulting from the recent East-West accords and the ending of the Cold War. It seems unlikely that production of M1A2 will stop at just 62 tanks, which is sufficient for just two battalions. Also, the US Army is known to have a new 140mm tank gun under development, which might, if suitable, be retrofitted to the existing fleet of M1s.

In addition to domestic production, 555 have been ordered by Egypt and there is potential for other orders, including, possibly, Saudi Arabia, Pakistan and Sweden. The M1-series has also competed with other modern MBTs for orders in the UK and Switzerland, but has failed to be selected.

Below: The sand billowing from its tracks and in its wake denote an M1A1 Abrams travelling at high speed through the desert. This elevated view clearly shows the distinctive angling of the turret's front and side plates, as well as the Rheinmetall M256 120mm smoothbore gun.

M41 Walker Bulldog
Light Tank

Country of origin: United States of America.
Crew: 4.
Armament: One 76mm gun; one 7.62mm machine-gun co-axial with main armament; one 12.7mm anti-aircraft machine-gun.
Armour: 0.36in-1.49in (9.25mm-38mm).
Dimensions: Length (gun forward) 26ft 11in (8.21m); length (hull) 19ft 1in (5.82m); width 10ft 6in (3.12m); height (including 12.7mm machine-gun) 10ft 1in (3.07m).
Weight: Combat 51,800lb (23,495kg).
Ground pressure: 10.24lb/in² (0.72kg/cm²).
Engine: Continental or Lycoming AOS-895-3 6-cylinder petrol engine developing 500bhp at 2,800rpm.
Performance: Road speed 45mph (72km/h); range 100 miles (161km); vertical obstacle 2ft 4in (0.71m); trench 6ft (1.83m); gradient 60 per cent.
History: Entered service with the US Army in 1951. Still in service with Brazil (300), Chile (60), Denmark (52), Dominican Republic (12), Guatemala (10), Somalia (10), Taiwan (675), Thailand (200) and Tunisia (10).

The standard light tank in use with the United States Army at the end of World War II was the M24 Chaffee, which weighed 40,285lb (18,289kg) and was armed with a 75mm gun. Shortly after the end of the war work started on a new light tank called the T37. The first prototype of this was completed in 1949 and was known as the T37 Phase I. This was followed by the T37 Phase II, which had a redesigned turret and different fire-control system. This model was then redesignated as the T42 and a slightly modified version of this, the T41E1, was standardised as the M41. The M41 was authorised for production in 1949 and was named the Little Bulldog, although the name was subsequently changed to the Walker Bulldog after General W. W. Walker, killed in an accident in Korea in 1951. Production of the M41 was undertaken by the Cadillac Car Division of the General Motors Corporation at the Cleveland Tank Plant, and first production models were completed in 1951. Further models of the M41 were the M41A1, M41A2 and the M41A3. These have a slightly different gun control system, whilst the M41A2 and M41A3 have a fuel-injection system for the engine. The M41, as one of the three main tanks developed for the US Army in the early 1950s, the others being the M47 medium and the M103 heavy tanks, was the first member of a whole family of vehicles sharing many common components. The family included the M42 self-propelled anti-aircraft gun, the M44 and M52 self-propelled howitzers and the M75 armoured personnel carrier. In addition there

Right: The M41 light tank was one of three tanks developed by the United States in the early 1950s, the others being the M47 medium tank and the M103 heavy tank. The M41 shares many common components with the M42 twin 40mm self-propelled anti-aircraft gun and the M44 (155mm) and M52 (105mm) self-propelled howitzers. It was replaced in the United States Army by the M551 Sheridan but large numbers of M41s remain in service with other countries in all parts of the world. Main armament consists of a 76mm gun.

were many trials versions in the 1950s. Some M42s were also used as targets by the United States Navy as the QM41. Fitted with remote-control equipment, they are used as mobile targets for new air-to-ground missiles. The hull of the M41 is of all-welded steel construction, whilst the turret is of welded and cast construction. The driver is seated at the front of the hull on the left, with the other three crew members in the turret, the commander and gunner on the right and the loader on the left. The engine and transmission are at the rear of the hull, and are separated from the fighting compartment by a fireproof bulkhead. Like most American AFVs of that period, the M41 is provided with a hull escape hatch, thus enabling the crew to leave the vehicle with a better chance of survival than if they baled out via the turret or driver's hatch. The suspension is of the torsion bar type and consists of five road wheels, with the drive sprocket at the rear and the idler at the front. There are three track return rollers. The main armament of the M41 consists of a 76mm gun with an elevation of $+19°$ and a depression of $-9°$, traverse being 360°. A 7.62mm machine-gun is mounted to the left of the main armament and there is a 12.7mm Browning machine-gun on the commander's cupola. Some 65 rounds of 76mm, 2,175 rounds of 12.7mm and 5,000 rounds of 7.62mm ammuntion are carried. The barrel of the 76mm gun is provided with a bore evacuator and a 'T' type blast-deflector, the latter's function being to reduce the effects of blast and obstruction caused by the flow of propellant gases into the atmosphere. These gases otherwise raise a dust cloud and make aiming of the weapon more difficult.

The M41 was replaced by the M551 Sheridan in US Army service and has also been discarded by a number of other armies. Well over 1,000 remain in service, however, and several armies, including those of Brazil, Taiwan and Denmark have been giving their fleets substantial upgrades to extend their service by at least another 10 years. Various upgrade packages are on offer to other users, most of which involve replacing the 76mm gun with a 90mm (or even 105mm) weapon and replacing the original six-cylinder petrol engine with a more fuel-efficient diesel powerplant.

M47 Medium Tank

Country of origin: United States of America.
Crew: 5.
Armament: One M36 90mm gun; one M1919A4E1 .3in machine-gun in bow; one M1919A4E1 .3in machine-gun co-axial with main armament one M2 .5in machine-gun on commander's cupola.
Armour: 0.50in-4.60in (12.7mm-115mm).
Dimensions: Length (gun forward) 28ft 1in (8.1m); length (hull) 20ft 8in (6.1m); width 10ft 6in (3.51m); height (including anti-aircraft machine-gun) 11ft (3.35m).
Weight: Combat 101,775lb (46,170kg).
Ground pressure: 13.3lb/in² (0.93kg/cm²).
Engine: Continental AV-1790-5B 12-cylinder air-cooled petrol engine developing 810bhp at 2,800rpm.
Performance: Road speed 30mph (48km/h); range 80 miles (130km); vertical obstacle 3ft (0.914m); trench 8ft 6in (2.59m); gradient 60 per cent.
History: Entered service with the US Army in 1952. Still in service with Greece (390), Iran (100), Italy (313), Jordan (260), Pakistan (300), Somalia (120), South Korea (300), Spain (375) and Turkey (523). (*Note that in most cases most, if not all, are in storage and not in front-line service*).

When the Korean War broke out the M26 and M46 were the standard US Army tanks, but a new medium tank was needed urgently. This was produced by mounting the turret of a new experimental tank, the T42, on the hull of the existing M26 Pershing. The new tank, which was considered an interim design by the US Army, became the M47 Patton and 8,676 were built between 1950 and 1953 — not a bad number for a "stop-gap"!

The hull and turret are of all-cast construction. The driver is seated at the front of the hull on the left. The other four members of the crew are the bow machine-gunner to the right of the driver, and the usual turret crew of commander, gunner and loader. The main armament is the M36 90mm gun, which has a T-shaped blast deflector and which fires a variety of rounds, including HEAT and HEAT (Fin-Stabilized). MECAR in Belgium and IMI in Israel have recently developed

Below: The M47 was developed during the Korean War and is essentially a modified M26 Pershing chassis fitted with the turret developed for the T42 tank. Over 8,500 M47s were completed in the 1950s.

90mm APFSDS rounds for this weapon. 71 rounds of 90mm ammunition are carried. There is a co-axial 7.62mm machine-gun, a 12.7mm anti-aircraft machine-gun, and a 7.62mm bow machine-gun.

The US Army relegated the M47 to reserve status after only a few years, but many were supplied to NATO countries, and, although few of these remain in front-line service, there are many in storage. A new tank factory was built in Iran between 1970 and 1972 and the first tank selected for production there was a developed version of M47 designated M47M. This retained the 90mm main gun, but had a number of components from the M48A3 and M60A1 series, including engine, transmission, electrics and optical equipment. This resulted in a much superior vehicle, with greatly extended range. Some 400 were produced, of which about 100 are believed to remain in service.

Above: An M47 moves up to the battlefield with infantry support. The M47 was soon replaced in the US Army by the much-improved M48, but large numbers of this reliable MBT are still assigned to the armies of nine nations.

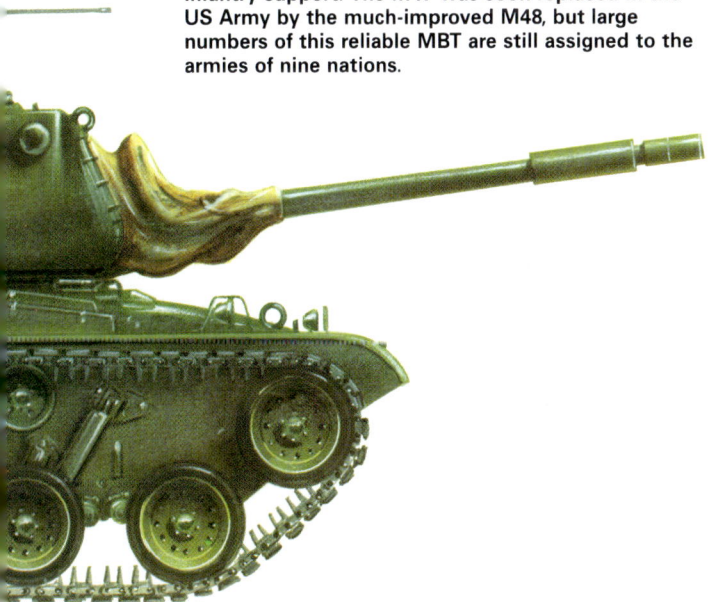

M48A5 Main Battle Tank

Country of origin: United States of America.
Crew: 4.
Armament: One M68 105mm rifled gun; one M60D 7.62mm coaxial machine-gun; two M60D 7.62mm anti-aircraft machine-guns.
Armour: Maximum 0.5in-4.8in (12.7mm-120mm).
Dimensions: Length (including main armament) 30ft 6in (9.31m); length (hull) 21ft 1in (6.42m); width 11ft 11in (3.63m); height 10ft 2in (3.09m).
Weight: Combat 107,900lb (48.987kg).
Ground pressure: 12.51lb/in² (0.88kg/cm²).
Engine: Continental AVDS-1790-2D. 12-cylinder, air-cooled, diesel engine developing 750hp at 2,400rpm.
Performance: Road speed 30mph (48km/h); range 310 miles (500km); vertical obstacle 3ft 0in (0.92m); trench 8ft 6in (2.6m), gradient 60 per cent.
History: Entered service with US Army in 1953. Now in service with: Germany (875), Greece (1,200), Iran (100), Israel (500), Jordan (c.200), South Korea (950), Lebanon (104), Morocco (173), Norway (38), Pakistan (345), Portugal (86), Spain (180), Taiwan (286), Thailand (175), Tunisia (28), Turkey (2,700), USA (1,013). *(Specification data apply to M48A5).*

A product of the rush to rearm consequent on the outbreak of the Korean War in 1950, the M48 design was completed in two months, prototypes were running a year later and the first production models were delivered in 1953 — a timetable that would be literally impossible today. The M48 was armed with a 90mm gun, had a crew of four and had a much-improved suspension compared to that of the M47. The original version was quickly followed by the M48A1 and M48A2, each of which had relatively minor improvements, and the first major advance was with the M48A3, with virtually all -A1s and -A2s in US Army service being upgraded to this standard. The M48A3 has a diesel engine replacing the petrol

Right: The M48A5 is an earlier M48 with many improvements, the most notable of which is the use of an M68 105mm rifled gun in place of the 90mm weapon. Some 8,950 M48s remain in use.

Above: M67A1 flamethrower tank in action in Vietnam. All US flamethrower tanks have been placed in storage for wartime use.

engine and improved fire control systems.

It was eventually realised that the day of the 90mm gun had passed, and it was decided to upgrade the M48 yet again by installing an M68 105mm main gun. This involved many other changes and produced an almost new tank, which was little different from the M60 (which itself had originally been simply an M48A3 with a 105mm gun).

The M48 was exported in very large numbers and some 8,950 remain in service with 17 armies. A small number of these are M48A2 or -A3s which retain their 90mm guns, but the vast majority have been upgraded to take the 105mm. Apart from the US Army's M48A5 conversions, one of the biggest of these programmes is that of Wegmann in Germany, which has converted 650 tanks to a new M48A2GA2 standard, with an L7A3 105mm rifled 105mm gun, and improved optical and fire control equipment. Other similar programmes have been or are being undertaken in Turkey, Iran and Israel.

M60A3 Main Battle Tank

Country of origin: United States of America.
Crew: 4.
Armament: One M68 105mm rifled gun; one M73 7.62mm co-axial machine-gun; one M85 12.7mm anti-aircraft machine-gun.
Armour: Classified.
Dimensions: Length (including main armament) 30ft 11in (9.44m); length (hull) 22ft 9in (6.95m); width 11ft 11in (3.63m); height 10ft 9in (3.27m).
Weight: Combat 107,900lb (48,987kg).
Ground pressure: 12.37lb/in² (0.87kg/cm²).
Engine: Continental AVDS-1790-2c, 12-cylinder, air-cooled, diesel engine developing 750hp at 2,400rpm.
Performance: Road speed 30mph (48km/h); range 300 miles (480km); vertical obstacle 3ft 0in (0.92m); trench 8ft 6in (2.6m), gradient 60 per cent.
History: Entered service with the US Army in 1960. Now in service with Austria (170), Bahrain (54), Egypt (753), Ethiopia (22), Greece (110), Iran (200), Israel (1,400), Italy (300), Jordan (218), Oman (6), Saudi Arabia (258), Sudan (20), Tunisia (54), USA (9,800) and Yemen (64).
(Specification data apply to M60A3).

Below: The commander and gunner of an M60 confer over a map while on a US Army exercise in northern Europe.

Above: A US Marine Corps M60A1 MBT moves inland after being landed during an amphibious exercise in Turkey.

▶ In the 1950s the standard tank of the United States Army was the M48. In 1957 an M48 series tank was fitted with a new engine for trials purposes and this was followed by another three prototypes in 1958. Late in 1958 it was decided to arm the new tank with the British 105mm L7 series gun, to be built in the United States under the designation M68. In 1959 the first production order for the new tank, now called the M60, was placed with Chrysler, and the type entered production at the Detroit Tank Arsenal in late 1959, with the first production tanks being completed the following year.

From late in 1962, the M60 was replaced in production by the M60A1, which has a number of improvements, the most important being the redesigned turret. The M60A1 has a turret and hull of all-cast construction. The driver is seated at the front of the hull with the other three crew members in the turret, commander and gunner on the right and the loader on the left. The engine and transmission are at the rear, the latter having one reverse and two forward ranges. The M60 has torsion-bar suspension and six road wheels, with the idler at the front and the drive sprocket at the rear; there are four track return rollers. The 105mm gun has an elevation of + 20° and a depression of − 10°, and traverse is 360°. Both elevation and traverse are powered. An M73 7.62mm machine-gun is mounted co-axially with the main armament and there is a 12.7mm M85 machine-gun in the commander's cupola. The latter can be aimed and fired from within the turret, and has an elevation of + 60° and a depression of − 15°. Some 60 rounds of 105mm, 900 rounds of 12.7mm and 5,950 rounds of 7.62mm ammunition are carried. Infra-red driving lights are fitted as standard and an infra-red/white light is mounted over the main armament. All M60s have an NBC system. The tank can also be fitted with a dozer blade on the front of the hull. The M60 can ford to a depth of 4ft (1.22m) without preparation or 8ft (2.44m) with the aid of a kit. For deep fording operations a schnorkel can be fitted, allowing the M60 to ford to a depth of 13ft 6in (4.114m). A radical departure was made with the M60A2, developed in the mid-1960s, in which a standard M60 hull was mated to a new turret mounting the (then) new 152mm gun/launcher. This could fire a Shillelagh missile or a variety of conventional rounds with combustible cartridge cases, a concept it shared with the M551 Sheridan light tank. This programme suffered endless problems and only 526 M60A2s were built before a halt was called. Those M60A2s which had been deployed to Germany were returned to the USA; later, the troubled tank was withdrawn from service.

Meanwhile, plans were being made to improve the 105mm gun armed version. A new fire control system, a laser rangefinder and a computer substantially enhanced the probability of a first-round hit, later helped even more by the addition of a Tank Thermal Sight. Many of these M60A3s came from new production, but others were upgraded M60A1s.

The Pentagon intends that the M60 will be phased out of the US inventory

Above: M60A2 MBTs during a night-firing exercise. The tank is armed with a 152mm gun/launcher which can fire a wide range of conventional types of ammunition or the Ford Shillelagh missile.

by 1997. Thus, any further improvements will be confined to those involving safety or protection of the environment.

Second-largest user of the M60 series is the Israeli Army, which has some 1,400 M60s, M60A1s and M60A3s, which are being constantly upgraded. All have been given more powerful versions of the Teledyne Continental diesel engine, while the locally-produced M68 105mm rifled gun has been fitted with an Israeli-developed thermal sleeve. Israeli M60s have also been fitted with Blazer explosive reactive armour, which consists of specially-tailored blocks bolted to the outside of the hull and turret, to give protection against chemical energy (CE) warheads. A newer add-on armour, called MAGACH-7, is now being fitted which is more bulky and substantially changes the appearance of the tank, particularly of the turret; this gives increased protection against both CE *and* kinetic energy projectiles. A new fire control system called MATADOR is also being installed.

Below: A total of 526 M60A2s were built at Detroit in the 1960s, but due to many problems the tank did not enter service with the US Army until the mid-1970s.

M107 175mm Self-propelled Gun/M110A2 203mm Howitzer

Country of origin: United States of America.
Crew: 5.
Armament: **M107** One 175mm gun; **M110A2** one 203mm howitzer.
Dimensions: Length (gun forwards) **M107** 36ft 11in (11.26m), **M110A2** 35ft 3in (10.73m); length (chassis) (**both**) 22ft 11in (7.0m); width (**both**) 10ft 10in (3.3m); height (to top of barrel when travelling) **M107** 12ft 1in (3.68m), **M110A2** 10ft 4in (3.14m).
Weight: **M107** 62,098lb (28,168kg), **M110A2** 62,500lb (28,350kg).
Engine: Detroit Diesel 8V-71T turbocharged 2-stroke liquid-cooled 8-cylinder diesel developing 405bhp at 2,300rpm.
Performance: Road speed **M107** 34.8mph (56km/h), **M110A** 34mph (54.7km/h); road range **M107** 450 miles (725km), **M110A2** 325 miles (523km); vertical obstacle (**both**) 2ft 11in (1.11m); trench **M107** 7ft 9in (2.36m) **M110A2** 6ft 3in (1.91m); gradient 60 per cent.
History: **M107** Production completed. In service with Greece (36), Iran (8), Israel (140), Italy (18), South Korea (numbers not known), Spain (12), Turkey (36). **M110** entered service in 1962. In service with Greece (31), Iran (38), Israel (48), South Korea (99) and Spain (12).

Above: The hydraulically-operated spade at the rear of the M110 is lowered before firing commences to provide a stable base.

M110A2 entered service in 1978. In service with Belgium (11), Germany (226), Italy (23), Japan (201), Jordan (1,012), Netherlands (76), Pakistan (40), Taiwan (60), Turkey (16) and USA (1,071).

In 1956 the United States Army issued a requirement for a range of self-propelled artillery which would be air-transportable. The Pacific Car and Foundry Company of Washington were awarded the development contract and from 1958 built three different self-propelled weapons on the same chassis. These were the T235 (175mm gun), which became the M107, the T236 (203mm howitzer), which became the M110, and the T245 (155mm gun), which was subsequently dropped from the range. These prototypes were powered by a petrol engine, but it was soon decided to replace this with a diesel engine as this could give the vehicles a much greater range of action. When fitted with a diesel engine the T235 became the T235E1 and after further trials this was placed in production as the M107 in 1962, entering service with the army the following year. The M107 has in fact been built by three different companies at various times: FMC, Bowen-McLaughlin

Above: M107 175mm gun in action during the Vietnam War firing an HE projectile weighing in at 151.5lb (68.78kg).

York and the Pacific Car and Foundry Company. It is not currently in production. The hull is of all-welded aluminium construction with the driver at the front on the left with the engine to his right. The gun is mounted towards the rear of the hull. The suspension is of the torsion-bar type and consists of five road wheels, with the fifth road wheel acting as the idler; the drive sprocket is at the front. Five crew are carried on the gun (driver, commander and three gun crew), the other eight crew members following in an M548 tracked vehicle (this is based on the M113 APC chassis), which also carries the ammunition, as only two ready rounds are carried on the M107 itself. The 175mm gun has an elevation of + 65° and a depression of − 2°, traverse being 30° left and 30° right. Elevation and traverse are both powered, although there are manual controls for use in an emergency. The M107 fires an HE round to a maximum range of 35,870 yards (32,800m). A large hydraulically-operated spade is mounted at the rear of the hull and is lowered into position before the gun opens fire, and the suspension can also be locked when the gun is fired to provide a more stable firing platform. The gun can officially fire one round per minute, but a well trained crew can fire ▶

▶ at least two rounds a minute. As the projectile is very heavy, an hydraulic hoist is provided to position the projectile on the ramming tray; the round is then pushed into the breech hydraulically before the charge is pushed home, the breech lock closed and the weapon is then fired. The M107 can ford streams to a maximum depth of 3ft 6in (1.06m) but has no amphibious capability. Infra-red driving lights are fitted as standard but the type does not have an NBC system.

The M110 203mm self-propelled howitzer is mounted on an identical hull to that of the M107 175mm and the barrels in both cases have the same elevation limits (+65° to −2°) and traverse (30° left and right of centre). The initial version of the M110, fielded in 1963, had a short, fat barrel, but this was replaced in the M110A1 with a much longer barrel which increased the range with a typical HE round from 18,373 yards (16,800m) to 25,044 yards (22,900m). The M110A1 was later upgraded to M110A2 standard, which basically involved the fitting of a double-baffle muzzle brake and the ability to use Charge 9, whereas the M110A1 was only able to use Charge 8. As listed above some armies have retained their basic M110s, but all those who took delivery of M110A1 have subsequently upgraded them to M110A2.

The M110A2 can fire a variety of projectiles. Prior to the withdrawal of chemical and nuclear artillery projectiles, the M110A2 could fire both. The nuclear round was the M422 with a maximum range of 19,800 yards (18,100m), which carried a W33 warhead with a yield of about 10Kt, which was about half that of the weapon dropped on Hiroshima. The chemical round (M426) could carry either Agent GB or VX. All that is now left for the M110A2 is the HE rounds, of which the most capable is the M650 HERA round, which has a maximum range of 32,810 yards (30,000m).

Above: An M110 of the British Royal Artillery shortly after being fired at the Royal School of Artillery, Larkhill, Wiltshire.

The M110 series have proved to be excellent weapons in service, although their lack of a turret to protect the crew would have been a major hazard in a nuclear war. Production ended in the early 1980s.

Above: An M107 opens fire in Vietnam. This self-propelled gun fires an HE round to a maximum range of 35,870 yards (32,800m).

Left: The M110 howitzer is mounted on the same hull as the M107 self-propelled gun, and also has the same elevation limits (+65° to −2°), with 30° traverse.

M109A2 155mm Self-propelled Howitzer

Country of origin: United States of America.
Crew: 5.
Armament: One 155mm gun; one 12.7mm or 7.62mm anti-aircraft machine-gun.
Dimensions: Length (gun forward) 29ft 11in (9.12m), length (chassis) 20ft 4in (6.2m); width (both) 10ft 6in (3.2m); height (to top of machine-gun) 10ft 10in (3.28m).
Weight: 55,000lb (24,948kg).
Engine: Detroit Diesel 8V-71T turbocharged 2-stroke liquid-cooled 8-cylinder diesel developing 405bhp at 2,300rpm.
Performance: Road speed 35mph (56.3km/h); road range 217 miles (349km); vertical obstacle 1ft 9in (0.53m); trench 6ft 0in (1.83m); gradient 60 per cent.
History: M109 entered production in 1962 and is still in production. In service with 26 armies: Austria, Belgium, Canada, Denmark, Egypt, Ethiopia, Germany, Greece, Iran, Israel, Italy, Jordan, Korea (South), Kuwait, Morocco, Netherlands, Norway, Pakistan, Peru, Portugal, Saudi Arabia, Spain, Switzerland, Taiwan, United Kingdom and the United States. The numbers produced as of 1992 are: **M109** — 3,194; **M109A1** — 3,418; **M109A2** — 2,457.

This outstanding design has been in continuous production for 30 years and is in service with 26 armies. It is by far the most widely used self-propelled howitzer in the world and is still being developed for service well into the 21st Century. It has a hull of all-welded aluminium construction, providing the crew with protection from small-arms fire. The driver is seated at the front of the hull on the left, with the engine to his right. The other five crew members are the commander, gunner and three ammunition members, all located in the turret at the rear of the hull. There is a large door in the rear of the hull for ammunition

Below: An M109A1 test-firing a Martin Marietta Copperhead Cannon-Launched Guided Projectile (CLGP) at the White Sands Missile Range, New Mexico, USA.

Above: A 155mm M109 self-propelled howitzer at a fire support base at Phu Bai in Vietnam. Well over 3,000 M109s have now been built.

resupply purposes. Hatches are also provided in the sides and rear of the turret. There are two hatches in the roof of the turret, the commander's hatch being on the right. A Browning 12.7mm machine-gun is mounted on this for anti-aircraft defence. The suspension is of the torsion-bar type and consists of seven road ▶

Below: The combat weight of the M109 has been kept to under 60,000lb (27,240kg) through the judicious use of aluminium in its overall construction.

Above: Prototype of the M109A1 was the M109A1E1 shown here at the US Army's Aberdeen Proving Grounds in Maryland. It can fire to a maximum range of 19,685 yards (18,000m).

▶ wheels, with the drive sprocket at the front and the idler at the rear, and there are no track return rollers.

The 155mm howitzer has an elevation of + 75° and a depression of – 3°, and the turret can be traversed through 360°. Elevation and traverse are powered, with manual controls for emergency use. The weapon can fire a variety of ammunition, including HE, tactical nuclear, illuminating, smoke and chemical rounds. A total of 28 rounds of separate-loading ammunition is carried, as well as 500 rounds of machine-gun ammunition. The second model to be introduced was the M109A1, identical with the M109 apart from its much longer barrel which is provided with a fume extractor as well as a muzzle-brake. The fume extractor removes propellant gases from the barrel after a round has been fired and thus prevents fumes from entering the fighting compartment. The M109 fires a round to a maximum range of 16,076 yards (14,700m), whilst the M109A1 fires to a maximum range of 19,685 yards (18,000m). The M109 can ford streams to a maximum depth of 5ft (1.83m). A special amphibious kit has been developed for the vehicle but this is not widely used. It consists of nine inflatable airbags, normally carried by a truck. Four of these are fitted to each side of the hull and the last to the front of the hull. The vehicle is then propelled in the water by its tracks at a maximum speed of 4mph (6.4km/h). The M109 is provided with infra-red driving lights and some vehicles have also been equipped with an NBC system.

Right: An M109A1 of the 7th United States Army based in Europe. It is deployed in battalions consisting of three batteries, each having six guns that can fire a tactical nuclear projectile.

M551 Sheridan Light Tank

Country of origin: United States of America.
Crew: 4.
Armament: One 152mm gun/missile launcher; one 7.62mm machine-gun co-axial with main armament; one 12.7mm anti-aircraft machine-gun; four smoke dischargers on each side of turret.
Armour: Classified.
Dimensions: Length 20ft 8in (6.30m); width 9ft 3in (2.82m); height (overall) 9ft 8in (2.95m).
Weight: Combat 34,898lbs (15,830kg).
Ground pressure: 6.97lb/in² (0.49kg/cm²).
Engine: Detroit Diesel 6V53T six-cylinder diesel developing 300bhp at 2,800rpm.
Performance: Road speed 45mph (70km/h); water speed 3.6mph (5.8km/h); range 373 miles (600km); vertical obstacle 2ft 9in (0.83m); trench 8ft 4in (2.54m); gradient 60 per cent.
History: Entered service with United States Army in 1966 and still in service with 82nd Airborne Division.

In August 1959 the United States Army established a requirement for a "new armoured vehicle with increased capabilities over any other weapon in its own inventory and that of any adversary". The following year the Allison Division of General Motors was awarded a contract to design a vehicle called the Armored Reconnaisance Airborne Assault Vehicle (ARAAV) to meet the requirement. The first prototype, designated XM551, was completed in 1962, and this was followed by a further 11 prototypes. Late in 1965, a production contract was awarded to Allison, and the first production vehicles were completed in 1966, these being known as the M551, or Sheridan. Production was completed in 1970 after 1,700 vehicles had been built.

The hull of the Sheridan is of all-aluminium construction whilst the turret is of welded steel. The driver is seated at the front of the hull and the other three crew members are in the turret, with the loader on the left and the gunner and commander on the right. The engine and transmission are at the rear of the hull.

The suspension is of the torsion-bar type and consists of five road wheels, with the drive sprocket at the rear and the idler at the front. There are no track-return rollers. The most interesting feature of the Sheridan is its armament system. This consists of a 152mm gun/launcher which has an elevation of $+19°$ and a depression of $-8°$, traverse being 360°. A 7.62mm machine-gun is mounted co-axially with the main armament and there is a 12.7mm Browning machine-gun on the commander's cupola. The latter cannot be aimed and fired from within the turret, and as a result of combat experience in Vietnam many vehicles have now been fitted with a shield for this weapon. The 152mm gun/launcher, a version of which was fitted to the M60A2 and MBT-70, fires a Shillelagh missile or a variety of conventional ammunition including HEAT-T-MP, WP and canister, all of them having a combustible cartridge case. The Shillelagh missile was developed by the United States Army Missile Command and the Philco-Ford Corporation, and has a maximum range of about 3,281 yards (3,000m). The missile is controlled by the gunner, who simply has to keep the cross-hairs of his sight on the target to ensure a hit. This missile itself weighs 59lbs (26.7kg) and has a single-stage solid-propellant motor which has a burn time of 1.18 seconds. Once the missile leaves the gun/missile-launcher, four fins at the rear of the missile unfold and it is guided to the target by a two-way infra-red command link which eliminates the need for the gunner to estimate the lead and range of the target. A Sheridan normally carries eight missiles and 20 rounds of ammunition, but this mix can be adjusted as required.

It was announced in 1978 that the M551 would be withdrawn from service, but a number remained operational in 1991 with the 82nd Airborne Division and took part in the liberation of Kuwait (Operation *Desert Storm*). Others remained in service in training roles at the National Training Center at Fort Irwin, California, some of which have been altered to resemble Soviet AFVs, and at Fort Knox, located in Kentucky.

Below: Courtesy of the 1.18 second burn time of its single-stage solid-propellant motor, a Shillelagh missile, its rear-mounted guidance fins beginning to deploy, streaks away from the 152mm gun/launcher atop an M551 Sheridan. A normal on-board complement is eight missiles per Sheridan.

Commando Stingray Light Tank

Country of origin: United States of America.
Crew: 4.
Armament: One Royal Ordnance LRF 105mm rifled gun; one 7.62mm M240 co-axial machine-gun; one M2 12.7mm anti-aircraft machine-gun.
Armor: Classified.
Dimensions: Length (including main armament) 30ft 6in (9.30m); length (hull) 21ft 2in (6.45m); width 8ft 11in (2,71m); height 8ft 3in (2.55m).
Weight: Combat 46,707lb (21,205kg).
Ground pressure: 10.24lb/in² (0.72kg/cm²).
Engine: Detroit-Diesel 8V-92 TA diesel developing 535hp at 2,300rpm.
Performance: Road speed 41mph (67km/h); range 300 miles (480km); vertical obstacle 2ft 6in (0.76m); trench 7ft 0in (2.13m); gradient 60 per cent.
History: First production Stingray completed in 1988; 108 now in service with Royal Thai Army.

For many years the world's major armies have concentrated on developing MBTs whose intended battlefield has been either the rolling plains of central Europe or the open deserts of the Middle East. As a result these MBTs have become ever larger, more sophisticated, more expensive and heavier, and armed with a gun of ever-increasing calibre. Some sales have been achieved outside these major theatres, but many Third World countries, who happily bought the relatively uncomplicated M47s or Centurions in the past, find these new MBTs increasingly unsuitable. The reasons can vary from sheer expense to something as simple as the facts that these heavyweight monsters are just too heavy for the local bridges.

There are, of course, a number of light tanks on the market, such as the British Scorpion family and the (albeit ageing) French AMX-13. However, many armies are looking for something in between, weighing about 30 tons (30.5 tonnes) and armed with a 105mm gun. Various tanks have been developed to meet this market. The British firm of Vickers developed a series of tanks, which have sold in some numbers to India, Kenya, Kuwait and Nigeria (*see Vickers MBT*), while Thyssen Henschel of Germany developed the TAM for the Argentine Army (*see pages 6-7*).

In the United States, Cadillac Gage also decided to target this market. After conducting some market research they set their priorities as: firepower (preferably a 105mm gun firing NATO-standard ammunition), high mobility, good operational range, minimum size, light weight, and transportability in a Lockheed C-130 Hercules aircraft. Conceptual work started in January 1983, design work in September 1983, with a prototype running in August 1984 and on demonstration

Below: A modern Light Tank of relatively straightforward design and construction, the Stingray totes an LRF 105mm rifled gun.

two months later. Cynics would say that such a very rapid development programme was only possible due to the total absence of government involvement, reinforced by the completely commercial pressures in the project up to this point! The prototype went to Thailand for trials in 1986 and the Royal Thai Army placed an order for 108 Stingray tanks in October 1987, which were delivered between 1988 and 1990.

With firepower and mobility given such high priority, protection is inevitably somewhat reduced. The hull is constructed of steel armour, which is capable of resisting penetration by a Soviet 7.62mm round over the entire vehicle, with increased protection at the front to resist a Soviet 14.5mm round. Since producing tanks to this specification for Thailand. Cadillac Gage has been investigating improvements to the armour protection to attract orders from other customers.

The gun is designed and produced by the British firm of Royal Ordnance and is developed from their very successful L7A3 weapon. The 105mm tube is fitted with a muzzle brake and a redesigned fume extractor, together with a thermal sleeve and muzzle reference system. The major change lies in the totally new recoil system, which reduces maximum trunnion force by a surprising 60 per cent to 30,000lb (13,608kg). 44 rounds of standard 105mm ammunition are carried, eight in the turret and the remainder below the turret ring in the hull. There is a coaxial 7.62mm machine-gun and either a 12.7mm or another 7.62mm machine-gun can be mounted on the turret roof above the commander's hatch. The Stingrays delivered to the Royal Thai Army are fitted with a British Marconi Digital Fire Control System, but a gun stabiliser is not fitted as standard, although this is an optional extra.

The Stingray is powered by the well-proven Detroit Diesel 8V-92 TA and the torsion bar suspension is based closely on that used for the past 25 years on the M109 155mm self-propelled howitzer.

Above: The sleek lines of the Stingray's turret are indicative of the manufacturer's aim to produce a tank with a low profile.

Below: To date, the sole customer for the Stingray has been the Royal Thai Army, with an order for 108 placed in October 1987.

OTHER SUPER-VALUE MILITARY GUIDES IN THIS SERIES

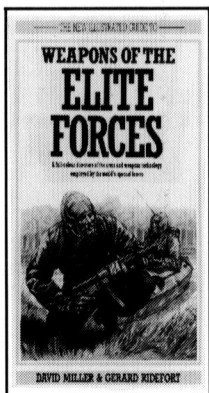

THE NEW ILLUSTRATED GUIDE TO

WEAPONS OF THE
ELITE
FORCES

DAVID MILLER & GERARD RIDEFORT

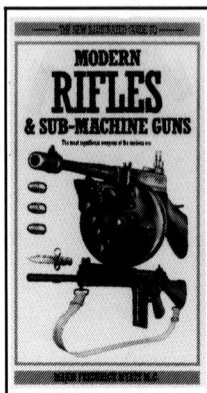

THE NEW ILLUSTRATED GUIDE TO

MODERN
RIFLES
& SUB-MACHINE GUNS

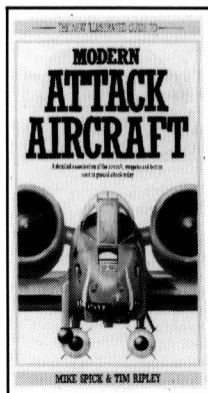

THE NEW ILLUSTRATED GUIDE TO

MODERN
ATTACK
AIRCRAFT

MIKE SPICK & TIM RIPLEY

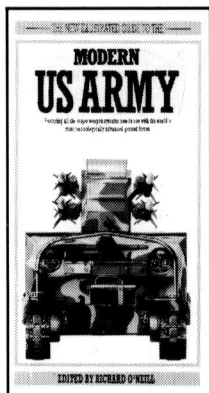

THE NEW ILLUSTRATED GUIDE TO

MODERN
US ARMY

EDITED BY RICHARD O'NEILL

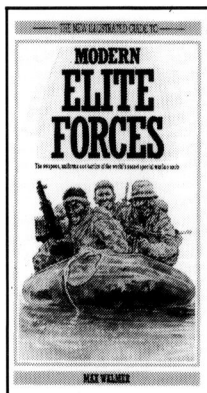

THE NEW ILLUSTRATED GUIDE TO

MODERN
ELITE
FORCES

MAX WALMER

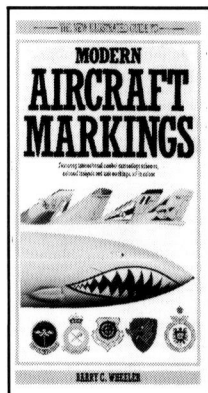

THE NEW ILLUSTRATED GUIDE TO

MODERN
AIRCRAFT
MARKINGS

BARRY C. WHEELER

OTHER ILLUSTRATED MILITARY GUIDES AVAILABLE.

Modern Warships
Modern US Navy
Modern Sub-Hunters
Modern US Fighters and Attack Aircraft
Allied Fighters of World War II

★ Each title has 160 fact-filled pages
★ Each is colourfully illustrated with hundreds of action photographs and technical drawings
★ Each contains concisely presented data and accurate descriptions of major international weapons systems
★ Each title represents tremendous value for money

If you would like further information on any of our titles please write to:
Publicity Department (Military Division), Salamander Books Ltd.,
129-137 York Way, London N7 9LG, United Kingdom.